The Peenemünde Wind Tunnels

The Peenemünde Wind Tunnels

A Memoir

Peter P. Wegener

Yale University Press   New Haven and London

Designed by James J. Johnson.
Set in Stemple Garamond by
Keystone Typesetting, Inc.,
Orwigsburg, Pennsylvania.
Printed in the United States of
America by BookCrafters,
Inc., Chelsea, Michigan.

*Library of Congress
Cataloging-in-Publication
Data*

Wegener, Peter P., 1917–
The Peenemünde wind
tunnels : a memoir /
Peter P. Wegener.
    p. cm.
Includes bibliographical
references and index.
ISBN 0-300-06367-9
(C : alk. paper)

1. V-2 rocket–History.
2. Germany. Heer. Heeresver-
suchstelle Peenemünde–
History. 3. Supersonic wind
tunnels–History.
4. Wegener, Peter P., 1917–
5. World War, 1939–1945–
Personal narratives, German.
6. World War, 1939–1945–
Germany–Peenemünde.
I. Title.
UF535.G3W43   1996
623.4'5195'0943–dc20
95-50289

A catalogue record for this
book is available from the
British Library.

The paper in this book meets
the guidelines for permanence
and durability of the Committee on Production Guidelines for Book Longevity of
the Council on Library
Resources.

10 9 8 7 6 5 4 3 2 1

For Annette and my three sons

All the glory of the world
Vanished under war and strife;
Yet to guard and to sustain
Is the highest prize in life.

—GOETHE
Insel-Verlag almanac for 1917

Contents

Acknowledgments

I am indebted to Henry Turner of Yale University for his sound advice. Katepalli Sreenivasan, also at Yale, read my manuscript and suggested improvements. My former wind-tunnel colleague Richard Lehnert permitted me to use his informal, unpublished notes (in German) on the fate of the wind tunnels. Siegfried Erdmann, the head of the basic research group and my first boss, provided information on the wind tunnels in the period before my arrival. For reasons explained in this book, he was unjustly forced to leave the Aerodynamics Institute in 1944, and he supplied me with much material on his fate. Ernst Stuhlinger, a leading physicist at the Peenemünde laboratories who, with Frederick I. Ordway III, recently wrote a monumental biography of von Braun, provided helpful hints.

David H. DeVorkin of the department of space history of the National Air and Space Museum kindly sent me a copy of the so-called Zwicky report, in which Fritz Zwicky, a professor of astrophysics at the California Institute of Technology, advised the U.S. Navy to move the wind tunnels to the United States. In addition, DeVorkin's own book gave me interesting insights. I am grateful to an anonymous reviewer who suggested adding radar to the atom bomb and the V2 as the third major new weapon system of World War II. Torsten Hess of the KZ-Gedenkstätte Mittelbau-Dora, a museum and memorial at the site of the underground factory of the V2 and the associated concentration camps at Nordhausen, which I recently visited twice, provided extensive unpublished information and photographs. On the same topic, Günther Gottman, the director of the Museum für Verkehr und Technik (Museum for Traffic and Technology) in Berlin, and his associate Holger Steinle shared their views on the disastrous conditions in the concentration camps and the V2 factory. Rainer Eisfeld of the University of

Osnabrück graciously provided archival material and recent studies of the Dora camp. During discussions with him in New York, he gave me an insight on his forthcoming book on this subject.

I am grateful to my friend Stephen J. Mayer, who thoughtfully edited the manuscript with advice extending beyond the mysteries of punctuation. Mafalda Stock did the final word processing. At Yale University Press, Jean Thomson Black provided encouragement and showed faith in the book, and Heidi Downey gave the manuscript a thoughtful final reading.

I am grateful for permission to use the illustrations, which follow page 000 and include maps and photographs of scenery and buildings, wind tunnels, models, and prototypes. Permission was granted by Deutsches Museum Bildstelle, Munich, Germany; KZ-Gedenkstätte Mittelbau-Dora, Nordhausen, Germany; and Dr. D. Sutcliffe, National Physical Laboratory, Teddington, England. Other illustrations are from the author's collection. The sources are abbreviated in the captions as DMB, Dora, NPL, and PW, respectively.

The Peenemünde Wind Tunnels

Introduction

I WAS BORN AND RAISED IN BERLIN. WHEN HITLER came to power in January 1933, I was a fifteen-year-old student at a private boarding school in the middle of Germany. After completing school and fulfilling the mandatory half-year of the *Arbeitsdienst,* the work service required of all young men, I matriculated at the University of Berlin in the fall of 1936 to study physics and geophysics. Here I did not follow the advice of my father, who had suggested that I first complete the required two years of military service, but because I was convinced that Hitler would launch a war sometime in the future, I applied for a deferment to take advantage of the remaining period of peace. In the fall of 1938, no further deferment was granted, and I had to report to an air force antiaircraft unit in a suburb of Berlin.

The German army occupied Czechoslovakia in March 1939, and during the night before the invasion we set up our guns in a potato field near Berlin. No enemy aircraft showed up. When Germany attacked Poland in September, we again defended Berlin without firing a shot. We spent the cold winter of 1939–1940 on Germany's western border, and in May we participated in the campaign against France. Late in the fall—long after the armistice—we left France, and after an interlude in Berlin during the winter of 1940–1941, when British airplanes did finally appear in small numbers, we were moved in June 1941 through Poland to the Russian border. From the first day of the German invasion of the Soviet Union, June 22, we accompanied the armored units that were assigned to take Moscow. At first we moved briskly, but my unit was decimated, and we got stuck near Smolensk while the main army kept moving toward Moscow. Until May 1943, when my story really starts, we remained in the area around the German-held city of Orel. During

1

the two years after we left Berlin in June 1941, I was twice granted a one-semester furlough to return to the university. I received a doctor of science degree at the end of the second leave in May 1943 and returned immediately to the eastern front.

The great break in my life occurred when, a short while after returning to the front, I was ordered to report to the Air Ministry in Berlin. There I was reassigned as a scientist to the supersonic wind tunnels of the army's Aerodynamics Institute, even though I had never heard the word *supersonic* in my life. The wind tunnels were part of the army's rocket laboratories near Peenemünde, a tiny fishing village on the coast of the Baltic Sea. The laboratories were founded in 1936 under the directorship of the twenty-five-year-old Wernher von Braun for the sole purpose of developing a long-range liquid-propellant missile carrying a warhead of explosives, later called the V2. The three years that started with my assignment to Peenemünde are the primary subject of my memoir. This period was of cardinal importance in my life; it saved me from a likely death in Russia, and it brought me in January 1946 to the United States, where I have stayed ever since.

A major British air raid on Peenemünde took place during the night of August 17, 1943, when I happened to be running the night shift at the wind tunnels. The Aerodynamics Institute escaped damage, and because the work on the aerodynamics of the V2 was practically completed, a decision was made to move the entire installation in steps to Bavaria. I was part of a skeleton crew that remained temporarily at the Baltic to complete certain experiments. I finally left for the south in the late spring of 1944. I had now spent about one year with the rocketeers, and my recollections of work, internal politics, people—including Wernher von Braun—and social life at the Peenemünde station cover this period.

The work in Bavaria, in contrast with our previous efforts, was relatively slow. We few soldiers were camouflaged as civilians—uniforms were hidden —and life was enjoyable, even though war was engulfing more and more of Germany. The scenery was beautiful, we experienced no air raids, and there was no fighting in our vicinity until the American army occupied our small town.

Shortly before the end of the war I was sent with an army officer assigned to the wind tunnels to retrieve the originals of the many reports written by the wind-tunnel staff. Those reports contained a unique body of knowledge on supersonic flow. The purpose of our journey was to preserve the documents for the period after the war. They had been stored for safe-keeping in a cave attached to the Mittelwerk, the underground factory in the Harz Mountains in central Germany where V2s were mass-produced. We drove to this place, made contact by chance with von Braun, who helped us get permission to enter the plant, found the documents, and returned to Bavaria. This short visit made me aware that inmates from a concentration camp were building the V2. I observed little of what went on in the tunnels of the Mittelwerk, and I did not see the concentration camp itself, but I could see and hear that the inhumane treatment of the prisoners was beyond anything I had heard about other industrial undertakings in Germany. With exceptions that I will cite, these facts have been glossed over in the postwar literature on the V2. There cannot be any doubt about the outstanding technological work done at Peenemünde in a remarkably short period of time. But the way the V2 was mass-produced, starting in the fall of 1943, negates much of the highly praised achievement that led directly to the U.S. space program. In the last chapter I look back on these matters from today's vantage point.

My fortuitous encounter with supersonic aerodynamics during the war resulted, in the fall of 1945, in an offer to move with the wind tunnels to the United States. I became a part of Project Paperclip, the operation to transplant German scientists to the United States, and what happened then closes this memoir.

In about 1980, I first considered putting on paper my thoughts about this part of my life; my friend Henry Turner, a historian at Yale University, suggested that I start by writing down everything I remembered without consulting other literature about the period. I followed his advice, and by 1982 I had finished a preliminary manuscript. After that, I searched quite successfully for letters, diaries, and photographs, and I contacted people who knew me during the period I was writing about. About two years ago I slowly started to work on the book at hand. In the process of writing I became aware of the fallibility of my memory, of which I will cite some examples, and have attempted to keep my own recollections separate from the mass of material now available.

ONE

From the Russian Front to Peenemünde

AT THE END OF MAY 1943, I HAD BEEN ON THE Russian front nearly two years. I was living in a dugout about 160 miles south of Moscow, near the village of Mtsensk, next to the railroad tracks between Orel and the Russian-held city of Tula. The front was generally quiet and the weather was not too bad, but the future was unclear.

I was a second lieutenant in an antiaircraft unit. Although the antiaircraft service was a part of the German air force, antiaircraft units had been found useful by the army as well. The 88-mm flak cannons were the most effective antitank weapons, in particular against the formidable Russian T-34. The 20-mm rapid-fire guns—some of which had four barrels on one pivot—were useful in attacking a variety of targets on the ground and were indispensable in the defense against the low-flying, partially armored Stormovik single-engine aircraft. Usually flak regiments or even smaller units down to single guns were assigned on a rotating basis to different types of ground troops, such as motorized infantry or armored units. Such changing assignments gave us an overview of the events of the war in one substantial sector of the front. From our observations of everyone from infantrymen to generals, and everything from horse-drawn wagons to tanks, we developed a sense of how things were really going.

One morning, in pouring rain, a messenger from a higher headquarters appeared at our outpost and handed me a telegram. I was to pick up orders to leave immediately for an assignment at the Flak Research Division of the Air Ministry in Berlin! First I danced around in the rain, then I pulled myself together and went back to the control center of my unit to turn over my responsibilities, hoping that nothing much would happen during my last minutes at the eastern front, within hailing distance of the Russian lines.

Thus I was fortunate to leave the Russian front and embark on a journey that would take me to the supersonic wind tunnels of the army research laboratory at Peenemünde. Why I was chosen to work at the wind tunnels is still not completely clear to me, but the fact that I had finished my formal studies was instrumental in this development.

After finishing high school in the spring of 1936, I had to serve half a year in the Arbeitsdienst before I could enroll at a university. I then applied for a delay of the required two-year service in the armed forces (*Wehrmacht*) so that I could begin to study physics and geophysics. The delay was granted, and I matriculated at Friedrich-Wilhelm-Universität of Berlin (today Humboldt University, in the eastern part of the formerly divided city).

In two academic years I took basic courses in physics, chemistry, and mathematics, and got a whiff of geophysics. More important, during the summers of 1937 and 1938, I joined two expeditions to the Arctic. These exciting ventures were started by a group loosely affiliated with my former school, a progressive boarding school in Holzminden on the Weser River, not far from Hamelin of Pied Piper fame. The expeditions were led by the chemist and mountaineer Herbert Rieche, and Leo Gburek, an experienced geophysicist from the University of Leipzig, joined us. Gburek was responsible for meteorology and oceanography, and I became his assistant. To spend several months each year in Spitsbergen, an archipelago far north of Norway and east of Greenland, was a dream come true for me at the age of nineteen. The eight members of the expedition represented several disciplines, and we camped on the shore of the Hornsund, the largest southern fjord of the main island of Spitsbergen (or Svalbard, to give the archipelago its proper Norwegian name).

Several large glaciers ended in the fjord, and we planned to study the interplay of melting ice from the glaciers and the outer seawater. We took water samples and recorded temperatures as a function of depth at many locations. Sitting in our small open boat surrounded by floating ice and waiting for the measuring instruments to adjust to a given depth, I played the harmonica to attract curious seals, who looked us over. In addition, we observed the weather, studied radiation and magnetic fields, climbed previously unconquered mountains—led by the experienced Austrian mountaineer Rudolf Bardodej—and skied pulling sleds to the southern tip of the island. We shot ducks for food, and had a strenuous but glorious time.

In the fall of 1938, following my fourth semester, my exemption from the draft ended. In November, I reported to a casern of antiaircraft units located in a suburb of Berlin. It is not easy to summarize in a few pages the next four and a half years, which were the most demanding of my life. Basic training followed by motorized field trips filled the time until Germany attacked Poland in September 1939. At that point my unit was moved overnight from firing practice at the Baltic—shooting at targets towed by airplanes—to the roofs of the Siemens works in Berlin. During the Polish campaign no enemy aircraft appeared over the capital. In October we took up positions in the west near the border with Luxembourg. There we froze in the unseasonably cold weather during what is now called the phony war.

In May 1940 the attack against France began. From the first day, we moved through Luxembourg and Belgium, two neutral countries that were not prepared to resist such attack. In France we drove—literally by day and night—to the English Channel near Dunkirk. We were not involved in any ground fighting, except for dodging the occasional artillery shell. It was our job to shoot down low-flying airplanes, mostly small British bombers, and we had only very mixed success. Next we turned south, and we reached the Swiss border when the armistice was called. My unit had few casualties other than those caused by accidents with motor vehicles.

A period of guarding airports in Normandy followed. Some of us (I was not included) practiced putting our smaller guns on rafts to be pulled across the channel for an invasion of Great Britain. Fortunately this foolhardy plan was called off, largely, I believe, because of the lack of ships of all sizes. In the fall my unit moved by train to Berlin to set up 20-mm guns on the rooftops of various important buildings. Nearly every night there was an air-raid alarm when a few British aircraft appeared over the city. Searchlights scanned the sky, and much firing of the 88-mm guns took place. I never saw a bomber being shot down at that time, and we never fired a single shot. The only danger we encountered was the rain of metal debris from exploded shells that had not hit a target.

By now I was a corporal, and I was ordered to attend officers' school in Bernau, a small town close to Berlin. This period was a great interlude. We had classes by day, and I spent many evenings and long weekends at my father's house, which was readily accessible by the fine electric train system. I do not recall that we learned much of value to equip us for a commanding role in a war of the extreme nature that we later encountered in the east. Fortunately, none of us had an inkling of what was to come.

In the spring of 1941, I was promoted to a rank equivalent to that of sergeant. During the fateful month of June 1941, my unit moved from Berlin to Poland. On June 22, the day of the German attack on the Soviet Union, I was sitting on the west bank of the Bug River north of the city of Brest-

Litovsk. This river formed the border of the Soviet Union, established in 1939 after the German attack on Poland, which was joined by the Soviets after they noted the rapid advance of the German army. As the attack on the Soviet Union began, we moved with the army that was supposed to conquer Moscow before autumn. But because of its heavy losses in July, my unit did not get far beyond the city of Smolensk. After being resupplied we were assigned to various divisions and corps that advanced or retreated without ever getting near Moscow.

My memory of the fighting in Russia has not dimmed much over the years. I was simply scared much of the time, although additional complex feelings, thoughts, and experiences occurred. They included curiosity about the actual course of daily events, attempts at seeing the broader picture, actions under pressure, relations with friends living and dead, and thoughts about what the future might bring. I had begun to like the Russian people, the scenery, and the cities and villages. I learned a little of the language, and all this made for a full life of a strange sort.

From the fall of 1941, when I was stationed in Orel, my diary—I kept diaries, violating all rules and regulations—yields the following information about my resumption of student life. On November 7 an ordinance was published allowing soldiers who had served for three years and had studied at least four semesters of science or engineering to apply for a study leave (*Studienurlaub*) of one semester. I quickly applied and equally quickly ran into trouble with the local commanders, who did not wish to lose an experienced survivor. The Russian winter, with its dramatic drop in temperatures, arrived at roughly that time, and I was told that I would become a second lieutenant in the reserve. The reserve designation applied because during three years of service I had steadfastly refused urging to become a professional officer. On November 15, I measured an outside temperature of $-15°C$ (5°F), a temperature at which one's hand sticks to metal. A few days later I heard through the grapevine that my leave had been approved by higher authorities, who had overruled my immediate supervisors. On November 23, I left for Berlin, where I arrived on November 26, and I matriculated at the university on December 8.

In Berlin I lived at home, went to the university daily, and enjoyed life. There were theaters, movies, and cabarets to visit, all functioning in spite of the air raids, which had not yet reached the intensity they reached later. At the start of my fifth semester I met Albert Defant, the oceanographer who headed the university's Institute of Oceanography and whose introductory lectures on fluid mechanics and oceanography I had already heard. Between the two wars Defant had been the guiding spirit of the *Meteor* expedition. The *Meteor* was an oceanographic research vessel that crisscrossed the

North and South Atlantic from 1925 to 1927, collecting data at many locations and discovering important geophysical features, such as the Mid-Atlantic Ridge. The Austrian Defant was known to be cool to the regime, to put it mildly, and at that stage he had very few doctoral students. I hoped that he would take me, which he did. It appears that the university administration kept him on only because of his towering international reputation in the field.[1]

During the 1941–1942 leave, word reached me that my friend Leo Gburek had died in action. He had been drafted by the military weather service and was assigned to duty as a meteorologist on routine data-gathering flights over the North Sea. On every trip the airplane descended once to the ocean surface to take a vertical survey of the atmosphere. This procedure had been observed, and his aircraft was shot down near sea level by a British fighter plane. On January 20, 1942, I traveled to the Institute of Geophysics at the University of Leipzig, where Leo and I had deposited our hydrographic and other geophysical results. I was given the material in the expectation that I would prepare it for publication. Extensive data reduction now lay ahead of me, since I had not seen some of the findings based on our raw data that were obtained in Leipzig after our return. For example, the salinity of the water in our sample bottles was determined at Leipzig, while we had measured temperature, oxygen content, and other properties in the field.

Thus without much warning I took responsibility for the analysis and publication of the geophysical results of our expeditions. I spoke with Defant and L. Möller, a professor at his institute who had taken a supportive interest in me. She helped patiently with the simple initial steps of data reduction that one would normally have learned in a more standard, lengthy course of study. The papers were deposited at the Berlin institute for safekeeping, and I returned to Russia on March 16. On the way I stopped in Warsaw to visit my closest friend from my earlier student days in Berlin. He had become a physician associated with a German military hospital for soldiers with head wounds. Medical students were normally drafted immediately after completing uninterrupted study for their M.D. degree, when they became officers in the medical service. In spite of his young age, my friend had to perform brain operations. Although by now I was used to the devastating effects of even the briefest military action, this visit to the hospital left a deep impression on me, since one did not usually witness the extended consequences of the human carnage.

Back on the front, we spent the summer pursuing various brief local campaigns of dubious value and defending Orel from the few Soviet airplanes that appeared by day. Any antiaircraft action during the night was futile. On September 15, I heard a rumor that gave me hope for a second study leave. I applied on October 12 but was turned down flat, since the need for healthy, experienced second lieutenants had grown in the past year. Nonetheless, I was due a regular leave because of the length of my service, and I went to Berlin for most of November. I found out then that only divisional commanders could turn down a study application. It seems that somebody had begun to recognize the need for technological developments, and my hope for a study leave was renewed. Indeed, early in January 1943, I traveled to Berlin for the third time in just over a year, this time to study again.

I went to the university and appeared at Defant's office. My adviser now made a most unexpected suggestion, based on his review of the oceanographic results of the Arctic expeditions. Considering the course of the war, he thought that another study leave for me beyond the current one was unlikely. For this reason, he said, I should write up my studies of the Hornsund Fjord, which he would accept as a doctoral dissertation. The only difficulty was that at least seven semesters of study were required for a doctoral degree. I had put in only six semesters, but Defant said he would apply to the dean of sciences for an exemption. The dean was Ludwig Bieberbach, a famous mathematician who, I was told, was also a fervent, idealistic follower of the current regime. It is probably difficult to understand the seemingly contradictory relation between science and politics unless one has lived under a totalitarian government. But Bieberbach was a decent man who disagreed with Defant's political views—I assume—but accepted his scientific judgment. He arranged for six semesters to be considered sufficient. If I could get the material together and pass the exams, I would receive the degree.

All these events looked too good to be true; indeed, my joy quickly came to an end. On February 18, I received orders to return immediately to my unit in Russia. My leave had all been a mistake, since in the air force study leaves were supposed to be granted only to students of medicine and pharmacy. Frantically, I ran to the Air Ministry, where I had met a friendly higher civil servant who understood that I did not intend to default on the service. Moreover, I visited my first regimental commander in Russia, who had retired and now lived in Berlin. I knew Colonel von Doering-Manteuffel from the first advance into Russia, when I had had (as a simple noncommissioned officer) discussions with him about how to distribute the effective flak guns among the advancing troops. He was a gentleman who personified the best Prussian tradition. He spoke quietly with all soldiers irrespective of rank, and he never greeted anyone with the standard, required "Heil Hitler." Now he apparently made phone calls, joining the chorus of my supporters. At any rate, at the last moment an order appeared in the ministry noting that the "research council of the *Reichsmarschall* [Göring] was interested in the completion of [my] examination in view of later assignments in research and development." This last turn of events—all reconstructed from letters that I wrote—strongly points to Kurt Wegener's intercession in my behalf, as I will explain in a moment.

To make a long story short, I completed my dissertation. I then passed a required exam in meteorology but flunked one in geography. I knew next to nothing in the latter field, and again Defant came to the rescue. He persuaded the geography professor to give me the lowest possible passing grade, protesting that I would surely never again receive a leave and that he believed I ought to be given a chance. Finally, on March 24, 1943, there followed my examination by Defant himself. I did a barely passable job, and at the end he told me that he expected I was aware of the fact that I knew next to nothing. I agreed, and he continued by saying that if I got through the war, he believed I would indeed make a scientist. (There was in fact one slight original contribution in the thesis, a suggestion of climatic development around Spitsbergen in thirty years obtained by comparison of our data with earlier work by the Arctic explorer Fridtjof Nansen.) As I left his office, thanking him for all he had done, Defant said he expected I knew that he had given me a political exam. I was perplexed until it dawned on me that professors were supposed to quiz a student not just on his field but also on his politics. At any rate, I had made it over this final academic hurdle, and on May 11, I received the degree of doctor of natural sciences.[2] Shortly thereafter, the events described at the beginning of this chapter took place.

My active contact with Defant was at an end. After the war, which he survived largely in Berlin, Defant became the *Rektor* (president) of the University of Innsbruck, in Austria. Some years later I sent him a number of reprints of papers of mine. Although the subjects were far removed from oceanography, I hoped they would tell him that his early trust in me was in some measure justified.

Before describing my journey to Peenemünde I want to recall what I believe to be the reasons for my assignment. At the time of my receipt of orders to report to the Air Ministry in Berlin, there was apparently taking place a concerted effort to collect surviving army, navy, and air force personnel with special skills in science and engineering. At roughly the same time there was also a personal intercession on my behalf by a first cousin of my father: Kurt Wegener, a professor of geophysics at the University of Graz, in Austria.

Like his brother Alfred of continental drift fame, Kurt was a balloonist and Arctic explorer.[3] The brothers made many notable balloon flights early in the century. On one occasion in April 1906 they remained aloft—though they were not prepared for the event—for fifty-two hours, beating the then-current endurance record by seventeen hours. This flying experience was noted, and Kurt joined a bombing squadron during World War I. His crew mate in one of the two-man planes was Hermann Göring, and he got to know Göring well.

Kurt Wegener had followed with disgust Hitler's rise to power, his government, and the events of World War II. In fact, he spent most of the war years in South America, bitterly and openly critical of the regime. Göring, of course, headed the German air force in addition to serving many other party and government functions. Early in 1943, Kurt wrote a strongly worded "Dear Hermann" letter to Göring saying that his nephew Peter had now served several years in the antiaircraft units of the air force, had participated in the campaigns in France and Russia at the front, had all kinds of decorations, and had also completed his doctoral work in physics and geophysics at the University of Berlin during two shortened semesters of furlough granted in 1941 and 1942. This letter was soon followed by the receipt in Russia of the telegram ordering me to appear in Berlin. Kurt wrote to me that he had got a prompt reply from Göring, who said he had turned over my case to the general in charge of the antiaircraft service. Whichever of the two avenues led to my transfer—or perhaps it was prompted by both—the result was that I entered, without any special preparation, the new world of supersonic aerodynamics.

My transfer from the front to Peenemünde began with the adventure of returning home to Berlin by various means. (I had done this on three previous occasions, a regular leave and two leaves for study.) A motorized carriage running on the rails connecting Orel and Tula took me back to Orel from Mtsensk, the current location of the front since all attempts to occupy Tula had failed. Orel was the city that I knew best and liked the most. I had spent much time there since the late fall of 1941. At this time the city was not as extensively damaged as many others, and consequently a movie house, an ornate turn-of-the-century bathhouse, and other amenities and recreational facilities were available. Orel was connected by train or aircraft with Smolensk, a beautiful city that I vividly remember from the first advance in 1941, prior to its destruction late in the war. In July of that year we took the city and were then surrounded by Russians. I recall that I listened on my shortwave radio to a Russian broadcast in German saying that the city had, in fact, been formally surrendered by its German occupiers. Since we were at the edge of town, at an airfield, we could not verify the broadcast, and we grew apprehensive. I saw no Russian soldiers except a pilot who innocently landed in a small airplane on the one runway. Fortunately for us, the surrender story was false.

But back to my final journey from the front. I flew to Smolensk in a half-empty Junkers 52 proceeding at treetop level because of extremely poor weather. At the airport I was told that a train to the west was to leave soon.

With cigarettes, I bribed the driver of one of the one-horse carriages to rush to the station. Indeed, a train carrying the wounded and soldiers on furlough stood ready to leave. This time I traveled a new route: instead of going through Brest-Litovsk on the border between the Soviet Union and Poland, we went north through the city of Vitebsk in partisan-held territory to Lithuania and from there to the German border—at that time East Prussia—which we reached after midnight on the second day. Here we underwent the usual delousing. I never got used to the strange procedure of passing naked through a sequence of showers and rubdowns with special ill-smelling fluids, while all one's belongings were gathered in a cloth bag and sent on a separate path to undergo other treatment. At any rate, at the end one felt refreshed. After we were reunited with clothing and baggage, a normal through-train (D-Zug) took us slowly to Berlin, and I vividly recall the enormous pleasure of arriving at my father's house in Binger Strasse in the western part of the city.

The next morning I put on my uniform and went to the Air Ministry. I was directed to a lieutenant colonel who turned out to be an Austrian schoolteacher. He interviewed me at length, asking in particular what I knew in the way of antiaircraft technology and science. I pointed out that I was thoroughly familiar with such antiaircraft weapons as the 20-mm and 88-mm guns, their associated mechanical computers, the optical equipment used to measure distances, and the theory behind firing on aircraft. At the time this theory was based on the assumption that an aircraft would proceed in a straight path at constant speed. A mechanical computer calculated the orientation of the artillery piece required to fire at a certain moment ahead of the aircraft on the assumed flight path, much as one would shoot at a running rabbit. This procedure was not successful if the pilot performed the slightest evasive maneuver. The 20-mm guns were in fact fired without use of their fancy but practically useless attachments. A single experienced person controlled the trajectory of the projectiles by observing their motion via illuminated tracers. Next I briefly mentioned my university studies.

The lieutenant colonel was not impressed by my knowledge, calling it "old stuff" with no use for the future. Soon aircraft would be shot down by entirely different means, he said. I found out much later that he had in mind electronically controlled, high-speed antiaircraft rockets—weapons that were far from completion even at the end of the war. In conclusion, he said that I would be assigned to the supersonic wind tunnels at the army's Peenemünde research establishment.

At the time of this interview, I had been in the air force for about four and a half years, time enough to become well versed in operating procedures. Therefore I simply answered "Jawohl" without letting on that I had never heard the word *supersonic*, not to mention *supersonic wind tunnels*.[4] I suppose that in the transcript of my university studies the term *fluid mechanics*—a field of which supersonic aerodynamics is a part—had appeared. I further suppose that nobody had noted the predominance of courses in oceanography, a science that includes the study of ocean currents, which move far more slowly than rockets. I felt relieved and happy when I left the Air Ministry, since it seemed to me that my career as a regular soldier was at an end. Indeed, this turned out to be true, except for a few memorable events involving confusion and danger that cannot be compared with what regular soldiers went through during the defeat and collapse of the German armed forces.

I received new travel orders and went by train to Stettin (now Szczecin, Poland). In retrospect, it is remarkable that in the early summer of 1943 the Allied air raids had not yet seriously affected people other than those in a few major cities. Trains ran on schedule, the mail service functioned well, people took vacations, and food—collected from all over Europe—was available in the amounts stated on ration cards. In fact, nothing approaching a "total war" existed even at that late stage, after the disaster of Stalingrad in January 1943, in which an entire German army was lost.

From Stettin, a local train running parallel to the Baltic coast brought me to Peenemünde, which is on the northwestern tip of a narrow peninsula called Usedom. The southeastern end of the peninsula is cut by the Swina, a small river that forms the current border between Germany and Poland, so Usedom is often called an island. As I entered the base, my papers were carefully checked by several uniformed men from units unknown to me—certainly not the regular armed forces. The army station covered a large open area that was fenced and heavily guarded. I received a temporary pass permitting me to move freely on the grounds. My orders were to join a unit called the *Flakversuchsstelle*, roughly "antiaircraft experiment station," which was commanded by a Captain König.

It was now late in the evening, and with some difficulty I found the barracks in which Captain König lived, and I rapped on his door. An older man in pajamas—old from my vantage point of twenty-five years—who was clearly in an unhappy mood, appeared at the door. I saluted and in a formal military voice declared my assignment to his unit. He pointed out that this introduction could have waited until the next morning and told me to find quarters for the night and come back tomorrow. This was the first indication that I would be fortunate to be encountering a military style different from the one I was used to. It turned out that König was a mathematician from the University of Göttingen and that the Flakversuchsstelle consisted of a motley crew collected from throughout the air force. The selection procedure had been based on presumed technical competence, and the civilian occupations of those chosen ranged from technician and craftsman to university professor in science or engineering. Military rank—from private to captain or major—played no role in this outfit. In fact, most us never even saw each other.

I soon learned that the base was devoted to the design, construction, and testing of liquid-fuel, long-range missiles that were to win the war for Germany. The facilities were operated by the army, with high-ranking officers formally in command. Rocket development had been pushed for years by Captain (later General) Walter Dornberger at army headquarters in Berlin and at outlying test sites before work began at Peenemünde in 1936.[5] Dornberger was joined by civilian rocket enthusiasts whose eyes were trained on the moon. A young engineering and physics student, Wernher von Braun, now known worldwide, began working with Dornberger in 1932. Von Braun later became the civilian technical director of the entire army research station, overseeing the laboratories and development facilities organized in many different subgroups and devoted to literally all aspects of rocketry.

The primary scientific and technical personnel at Peenemünde were civilians whose importance for the war effort kept them from being drafted. This group was augmented by technical specialists in the armed forces—people just like me—who worked exactly like the civilians. Fundamental research in applied mathematics on control systems, trajectories, and the like was carried on side by side with research on supersonic aerodynamics, combustion, chemistry, and structures, leading the way to such practical engineering projects as test-stand design and construction and experimentation with subcomponents and the entire rocket system itself. A liquid-oxygen production facility was built, prototype models were assembled in a pilot plant, and the design simplifications required for mass production were devised and tested. In short, Peenemünde represented one of the earliest examples of the design and production of what is now called a weapons system, down to the special trucks that carried the rockets. Although I lack special expertise in the history of science and technology, I think that the development of the atom bomb at Los Alamos, the development of radar in England and the United States, and the building at Peenemünde of the first large liquid-propellant rocket, the A4, are the first three examples of such broad undertakings.

The soldiers of the Flakversuchsstelle were distributed among the groups at the Peenemünde laboratories. Captain König told me that I would indeed be assigned to the supersonic wind tunnels, and he told me to check in with Rudolf Hermann, the director of the Aerodynamics Institute. By now I was tired, and based on my experience with König, I knew that it would be ill advised to search the station for the wind tunnels and Hermann, who was probably asleep. I postponed all further action, but I have no recollection of where I spent my first night at Peenemünde.

T W O

First Impressions

ON MY SECOND DAY IN PEENEMÜNDE, I COM-
pleted military formalities and found my way to the fascinating modernistic structure housing the wind tunnels. I introduced myself to Rudolf Hermann, who was expecting me. He
struck me as a person whose outgoing manner and appearance
were typical of the Wandervogel, a hiking, camping, and anti-city-life movement started in Germany at the turn of the century.[1] I next met Richard
Lehnert, the relaxed and friendly head of aerodynamics testing. Lehnert
asked about my background, and I felt secure enough to reveal my igno-
rance of aerodynamics. He took this confession quietly, suggesting that I
start reading Ludwig Prandtl's famous book *Essentials of Fluid Mechanics*.
He showed me to a pleasant, functional office, which I was to share with the
great physicist Pascual Jordan, who was attached to the wind tunnels in a
somewhat nebulous capacity, holding a rank and wearing a uniform of the
military civil service equivalent to that of an officer.

I was assigned to work not in the aerodynamics group but in the basic
research group under Siegfried Erdmann. To complete my formal absorp-
tion into the huge research center, I received quarters in Haus 30, a building
in the settlement (*Siedlung*) that had comfortable rooms for unmarried
officers and—more frequently—civilians who worked in the upper echelon
at the research center.

We ate in a mess hall shared with civilian employees and officers living
elsewhere. The food was good, though what one got was based on one's
ration cards, which were identical to those of the general population. There
was an occasional extra dish of fresh fish or some other seafood, but the total
caloric intake was equivalent to that of those members of the civilian popu-

lation—mostly city dwellers—who did not trade on the black market or have connections to farmers. The rations were much smaller than those I was used to as a soldier on the eastern front, and I quickly became even thinner than before, though I felt generally fine.

I now divided my time between the study of aerodynamics—the book-learning part of it—and work at the wind tunnels. I took every free minute to form a picture of where I was and what was being done at Peenemünde. The Army Research Establishment was at the northwest end of Usedom, on a strip of land bounded by the Baltic to the northeast and a shallow lakelike sound toward the mainland. The tip of the peninsula was occupied by an air force research and development laboratory, from which the army establishment was separated by a tight fence and stringent regulations. In fact, none of us were ever permitted to visit the lab.

The location on the Baltic Sea was ideal for secret rocket research in a country without open spaces like the deserts of the western United States. The test launching of rockets or unmanned aircraft under development by the army and the air force, respectively, could be carried out safely over the open sea. Usedom was suggested by the mother of Wernher von Braun because the family had spent summers at the resorts on the peninsula. A small island called the Oie served as the launching site for experimental army rockets even before the founding of the Peenemünde laboratories.

The strip of land lent itself perfectly to separation from the remainder of the country without a need for extensive fences. The space was open, with the sparse fir trees characteristic of the coast, some low ground cover, and much sand. The seashore provided a natural boundary for the research stations. Administration buildings, laboratories, housing units, and test stands were widely separated according to a well-planned layout, and a bucolic atmosphere prevailed. The architecture was attractive, combining a resemblance to the older municipal buildings of the northern provinces of Germany with a touch of the twentieth-century Bauhaus school. In contrast to most other architecture during Hitler's time, as exemplified by the work of Ludwig Troost and Albert Speer, the modern style had not been completely abandoned here. The army had put up simple and functional contemporary buildings for the laboratories. One concession to the style of the older local structures was the sloping roofs on buildings serving as living quarters, guardhouses, and the like.

The area instantly aroused nostalgia in me. Most of my early summers were spent at one of the many resorts on Usedom. For a child from Berlin, the Baltic seashore—with its pure white sand, its dunes, and its hunting grounds for shells and amber—was the closest ocean holiday spot. Once, to overcome a bout of whooping cough, I stayed until late fall, long after the other children had left for school. The scenery, the smell, and the waters of Peenemünde were truly familiar.

During one of my initial walks through the fir woods and along the beaches, among the rocket test stands, warehouses, and laboratories, I came upon an arenalike structure. In the center of this stadium, a rocket with fins and control surfaces at the rear was erected vertically and attached by bunched wires and hoses to a crane next to it. I had never seen such a strange object. I later understood that it was an Aggregat 4—called A4—an experimental guided missile. I have a distinct recollection that during this walk I observed a launching of this rocket; it rose slowly, then bent toward the Baltic to disappear in clouds at a great height. I remember an excruciatingly loud initial bang of the rocket engine, a screeching noise, clouds of deflected exhaust, into which a deluge of water was shot, and a reddish jet issuing from the back.

On reflection, it seems unlikely that I could have come upon a launching unexpectedly. There must have been some warning of an imminent test firing and more rigid control of access. But once inside the fence of the installation, one could move freely. My recollection precedes all later observations, viewing of films, discussions, and impressions of tests seen from a distance, and thus I believe that it is valid. Unfortunately, much of my 1943 diary has disappeared, breaking the flow of information about my experiences as a soldier. Here I must rely on—aside from too-often uncertain memory—letters, entries in notebooks, recollections of friends, and the like.

I was told that the A4 had at that time just begun to function reasonably well, reaching speeds of more than four times the speed of sound. It was later renamed the V2, with "V" for *Vergeltung* (retaliation), in response to the so-called terror attacks by Allied bombers that began to reach high levels in 1943. The name, like that of the air force's V1, was most likely coined by Goebbels's Propaganda Ministry at the time of the actual deployment of the weapons late in 1944.

The A4 was the latest in a series of liquid-propellant missiles developed by the army ordnance, with the first rockets predating not only the Peenemünde establishment but also Hitler's ascent to power in January 1933. The unusually brief period of development of the A4—about six years—culminated in the first successful firing on October 3, 1942, when a distance of 190 kilometers (118 miles) was covered. The fourteen-meter-long rocket, which had a starting weight of more than 12 metric tons (26,000 pounds, of which about 70 percent was fuel), reached an altitude of 85 kilometers (53 miles) and a speed of roughly 4,800 kilometers per hour (3,000 mph) during a five-minute flight. It could eventually cover a range of 290 to 340 kilometers (180 to 210 miles) carrying a 1,700-pound (or more) warhead of conventional explosives. Its guidance to a preselected target depended on attitude and location at the moment of a carefully timed fuel shutoff, or *Brennschluss*. Up to this moment, the flight path was controlled by four graphite vanes in the rocket jet's exhaust nozzle and small aerodynamic control surfaces—rudders—mounted on the edge of large external stabilizing fins.

The brilliant seminal technology of this first successful large-scale rocket has been discussed in innumerable general articles and books in addition to the technical literature, and the missile or models of it can be seen in science museums. Historical remarks and references to work related to this narrative will appear in chapter 4. The launching that I saw resembled that of present-day rockets, including the space shuttle. Such launches are by now familiar to every child via television, and it is remarkable that only about fifty years have passed since the days of Peenemünde.

A special electric rail system ran on the army grounds and connected with the regular railroad line on the peninsula. The impressive modern cars, patterned after those on the then-recent *Stadtbahn*—or above-ground urban trains of Berlin that augmented the underground trains—provided a convenient link with the outside world.

Another strong early recollection concerns the almost nightly buzzing noise caused by a flying bomb (designated Fi103 after the designer Fieseler) with the code name *Kirschkern* (cherry pit). This small unmanned aircraft, the later V1, was the antecedent of today's cruise missile, flying at about the speed of a British fighter plane. It was being developed as a long-range attack weapon in the air force laboratory at the northwestern tip of the peninsula. Propulsion was provided by an ingeniously designed engine. The cyclic shutting down of the combustion chamber by a valve resembling a venetian blind created a sequence of explosions of the fuel mixed with the air taken in through a pipe open in the direction of flight. All this led to the eerie buzzing. Since the test flights took place at night at low altitude on a flight path at a right angle to the coastline, away from land, one could peer into the exhaust and see, for a few minutes or so, a red dot—the cherry pit—tracing the missile's trajectory.

The completion of the development of the V1 and its use as a weapon preceded that of the A4, which employed a more radical technology. Except for its ingenious power plant, the V1 was based on long-standing principles of aircraft design, and so it was the first German "miracle weapon" to be used against Britain. My older friends in London who experienced the war tell me that the noisy V1—called the "doodlebug"—was more feared than the V2. It signaled its arrival like an airplane, and at the time the noise stopped, it started its silent descent. One knew that it was coming down; the resulting state of agonizing anticipation generated more fear than the sudden impact of a supersonic V2, whose approach was silent. An object moving at supersonic speed outraces its own sound.

All in all, I recall a rather pleasant life during my early weeks in Peenemünde. Moreover, I was continually mindful of the great advantage of not being involved in further fighting in Russia, fighting that turned increasingly into disaster for the German troops. I shared my father's frequently repeated view, based on his World War I experience, that in war, any place where nobody shoots at you is fine. I had no responsibilities for others or daily worries about survival. Moreover, I was learning a great deal of fascinating science and engineering and was slowly adapting to intellectual challenges.

There was free time in Peenemünde in spite of the heavy work load. Summer days invited walks in the countryside. The accommodations were pleasant, and in Haus 30 I met a number of interesting people, some of whom became my friends. The frequent nightly siren blasts signaling approaching aircraft did not concern us much. These air-raid warnings were usually triggered by overflights of large groups of British bombers moving toward Berlin. Occasionally a Mosquito—the famous British wooden reconnaissance airplane—appeared during the day at high altitude. Oddly enough, it never occurred to me that I lived in a most attractive location for an enemy air raid.

In Peenemünde, a regular life-style took hold: one could take showers daily, and the food, as I mentioned before, was good if not plentiful. I learned a great deal about the war, general technology, and politics (if one could call it that during Hitler's reign). There was permission to leave the research station, and I could walk or take the train to the many small resorts on the seashore that I knew so well from earlier days. The beaches were full of sunbathers, including many children. If one overlooked the relative lack of young men, things seemed much as I remembered them in peacetime. I also recall the weather in the summer of 1943 as particularly beautiful, in part, no doubt, because I was struck by my new-found security and freedom.

While I was involved in learning experimental supersonic aerodynamics, it did not dawn on me that the work being done here was unique. At the same time, the often-disastrous events of war at the front and in the large cities seemed strangely remote. To recap my first impressions, I cannot do better than to quote from one of my letters that fortunately survived the war. Early in June 1943, I wrote to my older brother in Berlin as follows (in translation, of course):

After appearing at the new place of assignment [I could not mention the Air Ministry in Berlin; a letter might be opened by censors, possibly leading to dire consequences], I was told yes, an experimental physicist was really needed; off to the northwestern tip of Usedom! There I was picked up by a car, and I was truly astounded by the inconceivable extent of all the facilities. At this place, members of the army and the air force, civilian scientists and girls are mixed together. I was assigned to the Aerodynamics Institute. All the supervisors are civilians, professors, and the like. The very well known theoretical physicists Jordan and Wessel are also here; what a circle, what work, what stimulation! For fourteen days now I have worked intensely, and tomorrow I do my first experimental shift [in the supersonic wind tunnel]. All this appears like the vision a small child may have of a researcher. Huge facilities, cranes and bridges, instruments, film, photography, electronic tubes, and everything in large numbers. I am responsible for two test engineers; the results are recorded by girls with long curls, mechanics. An office with a drafting board and all conceivable materials, completely furnished living quarters, and a great *Kasino* [the dining room for the civilian scientists and the officers]. Added to this, the sea is in front of us, and the only military duty is to eat lunch. Orderlies clean up. Can you still imagine how it is to run into the waves?

The counterbalance is the relatively poor accessibility of Berlin, complete solitude, the fact that Zinnowitz [a seaside resort where we often spent our summers] is off limits, and the long winter at the sea—all things that cannot scare me yet. [In retrospect, I have trouble seeing how I could have worried about the winter at Peenemünde, having escaped the Russian winter!] We work ten hours every day; our research is very interesting. A first lieutenant from the army who is an engineer [Günther Hermann, no relation to the head of the Aerodynamics Institute] is with us, the only officer on the staff, and we like each other.

Friday to Wednesday I go to Berlin as a courier by car and by train in a special compartment that keeps the masses of travelers away. . . . In the fourth year of the war that I have survived without being wounded, a new leaf has been turned over; since I am superstitious, I think about a possible reversal of the current happy circumstances. I will be able to do the work here; my previous education is not quite as poor as I always thought. I learn a great deal, in part because of the tension of the time. Therefore I learn thermodynamics, experimentation with many techniques, and aerodynamics, all directly applicable to my profession.

The thoughts and the mood expressed here were echoed in other letters. In spite of the note of secrecy in the first sentence, I find the description of my work remarkably open. Indeed, the security at Peenemünde was minimal in comparison with my later experience with classified materials at a U.S. military laboratory. People like me were not checked. No forms had to be filled out, and no questioning took place. Except on entering and leaving the main gate, no pass had to be shown anywhere, and no badges were worn.

Viewing my delight about the change of my circumstances, one might wonder why no thoughts appear on the implications of the new work. After all, I had entered a secret factory where wonder weapons were to be produced to win the war for Hitler. (Actually it took some time before I understood what these wonder weapons were.) Such complex ideas—ignoring the fact that I would not have dared to put them in a letter—did not enter my head at the time in view of the overwhelming relief at being away from the front and the joy of finally learning something about science in a laboratory setting. I will say more on this subject later.

T H R E E

The Supersonic Wind Tunnels

MONG THE TECHNOLOGICAL CONTRIBUTIONS
of many different disciplines to the development of a missile system, the external aerodynamic shape is the most visible. A rocket
must have a low air drag, it must be stable, and it must fly on a
controllable, predetermined course. In addition, a rocket needs to
depart from the purely ballistic trajectory of a thrown rock, an arrow, or an
artillery shell fired from a cannon. Stability of flight in the denser layers of
the atmosphere is provided by fixed external winglike surfaces and movable
rudders. Missiles with such aerodynamic controls were foreseen by the
pioneers of modern rocketry, most prominent among them Konstantin
Tsiolkovsky, Robert Goddard, and Hermann Oberth. Aerodynamic shapes
appeared even earlier in fictional accounts of space flight, such as those by
Jules Verne (whose moon rocket shows the low-drag shape of an artillery
shell), and in the classic 1920s science fiction film *Frau im Mond* (Woman in
the Moon) by the German director Fritz Lang, for which Oberth acted as a
consultant.

Despite these early intimations, at the time of the development of the A4
there existed no systematic body of experiments on missile aerodynamics.
This was particularly true for supersonic flight: experimental supersonic
aerodynamics was in its infancy. For this reason, it was essential to have
wind tunnels in which models could be tested at speeds above that of sound,
and such novel facilities were designed and constructed at Peenemünde.
There were two large supersonic wind tunnels with cross sections of 40 by
40 centimeters (16 by 16 inches) at the test section, the part of the tunnel in which air blasts of the required speed are produced and models are
mounted. An additional smaller research tunnel was also constructed. I was

told at the time that these wind tunnels were much larger than those at existing facilities. In a supersonic wind tunnel, the Mach number in the test section has to exceed the value of one. This number—now known to every newspaper reader—measures flow speed in a wind tunnel or, in turn, the flight speed of an object such as an airplane in units of the speed of sound. It is named after one of the earliest innovators in the science of high-speed flow, the Austrian physicist and philosopher Ernst Mach, who worked at the turn of the century.[1]

The immediate forerunners of the Peenemünde tunnels in Germany were those at Aachen. Carl Wieselsberger, the head of the Institute of Fluid Dynamics at the Technical University of Aachen, built in 1934 a 10-by-10-centimeter supersonic tunnel, in which much pioneering work was done by his assistant Rudolf Hermann. In 1936 the first aerodynamics tests were performed in this tunnel on models of the A3, the antecedent of the A4. Hermann and others were already working at that time on the design for the much larger Peenemünde wind tunnels. In 1937, one year after work started in the army laboratories at Peenemünde, the Aerodynamics Institute was founded, with Hermann as the director and with an initial staff of sixteen. In 1939, one of the large test sections began operating. At the same time, an electromechanical balance system was completed to determine the air drag and lift of models exposed to supersonic flows. The staff had grown to about sixty people, and research went remarkably well.

Of course, no technology springs up without prior advances in a field. In the nineteenth century, the Swedish engineer Carl de Laval had devised a nozzle to direct the steam in a turbine to the impeller blades that drive the turbine's shaft. He invented a converging-diverging passage that produced a supersonic jet at the nozzle's exit. As early as 1907, Ludwig Prandtl, one of the founders of modern fluid dynamics, published systematic studies of a steam jet operating at a Mach number of 1.5. He was also working in a new field called gasdynamics, the area of fluid dynamics dealing with high speeds, and he built a small supersonic research tunnel operating with air.[2] This small tunnel at Göttingen was preceded by a truly miniature facility at the National Physical Laboratory in England, where the drag of an artillery shell was measured in 1917 with a model the size of a pencil tip in a test section whose cross section was about the size of a finger.

Supersonic tunnels approaching the scale of those at Peenemünde appeared in the 1930s. One such facility was built near Brunswick by Adolf Busemann, who made notable contributions to theoretical gasdynamics. For example, Busemann, followed by Albert Betz, suggested in 1935 that the air drag of an airplane whose flight speed is close to the speed of sound could be reduced through the use of a swept-back wing, the characteristic

V-shape seen today on all large airliners. But the Brunswick wind tunnels never attained the operational proficiency, instrumentation, and staff size of the Peenemünde institute during the war. A similarly large tunnel was designed at the Italian research center Guidonia near Modena. Finally, under the direction of Jakob Ackeret, a Swiss pioneer of gasdynamics, an advanced supersonic wind tunnel was built in Zürich at the Swiss Technical University (Eidgenössische Technische Hochschule). Yet none of these facilities remotely equaled those of Peenemünde, a fact surely determined in part by the seemingly limitless supply of funding for the development of the rockets whose aerodynamics were being tested in the Peenemünde tunnels. Strangely, no similar facilities existed in the United States before World War II, though America now has the most extensive research apparatus in gasdynamics anywhere.

To move air continuously at supersonic speeds in a closed circular duct requires compressors of great power. In order to diminish the amount of power needed, the Peenemünde tunnels were based on another fundamental idea contributed by Prandtl, who had tried—on a small scale—a system of intermittent operation for his research tunnel. At Peenemünde, a huge riveted steel sphere with a diameter of more than 12 meters (40 feet) and a capacity of about 1,000 cubic meters was evacuated by three large pairs of rotary vacuum pumps driven by electric motors with a total power of 800 kilowatts. Once a certain low pressure was attained in the sphere, an ingeniously designed valve opened a pipe leading to the wind tunnel. For periods of roughly ten to forty seconds, with the longer testing period associated with the higher Mach numbers, air rushed through the test section into the vacuum sphere. Once the pressure in the sphere became too high to sustain the flow, the valve was closed again.

The air entering the wind tunnel was taken from outside, and before passing into the inlet of the test section it moved at low speed through an extensive drying system. Here the normal atmospheric moisture was absorbed by a hygroscopic substance called silica gel. The periodic drying of this absorber by heating it electrically took another 600 kilowatts or so. Drying was important, because in the nozzle and the test section the air expanded and attained low pressures and temperatures while reaching supersonic speed. If humid air was expanded in the nozzle, the water vapor in it might condense, just as it does in a cloud. The resulting droplets in the flow would disturb aerodynamic measurements in the test section, so they must be prevented from forming.

A direct link from the discovery of the condensation problem led to this unique application of an air drying system. The first international meeting on high-speed flow including gasdynamics, the Volta Congress, was held in

Rome in 1935. As an indication of the rapid growth of this new field, the congress was the first and last international gathering attended by literally all of the high-speed aerodynamics researchers in the world. Of the roughly thirty experts in fluid dynamics that were gathered, at most twenty were familiar with supersonic flow. At the Volta Congress, Prandtl reported on strange shocklike phenomena that he had observed in his supersonic nozzle. Wieselsberger, Hermann's mentor, suggested in the discussion that this trouble might arise from condensed water vapor if atmospheric air was used to operate a supersonic wind tunnel. Back in Aachen, he suggested to Hermann that he undertake a systematic study of the problem. Hermann then proved experimentally that water condensation was indeed at the bottom of this strange effect. He could relate the existence and strength of the disturbances in the flow to the relative humidity of the air entering the nozzle. These results led to the installation of the drying system at Peenemünde, a system that was not applied elsewhere in small or large tunnels and was a major reason that the aerodynamics facilities at Peenemünde were far ahead of their time.

The sequence of steps in performing an aerodynamics test was roughly as follows: The tunnel operator mounted a model of a given flying object on a support in the test section. Often the model was attached to the external electromechanical balance mentioned above, by which aerodynamic forces could be measured directly. The attitude, or angle of attack, of the model with respect to the airflow could be changed while the tunnel was in operation. Forces such as drag (the air resistance encountered by the model) and lift (the force acting at a right angle to the direction of the wind) could then be measured. From well-established laws of aerodynamics, the so-called fundamental similarity laws, the forces acting on the full-size prototype could next be computed based on the (much smaller) forces measured for the model. Various other pressure and electrical gauges attached to the model could provide additional measurements of various kinds.

Access to the model was obtained by moving two sets of thick framed plate-glass side windows mounted on rails on each side of the tunnel. In addition, an inner pair of windows enclosed the nozzle. Since the pressure in the test section was low during an experiment, these outer windows were exposed to large forces from external atmospheric pressure. Once, while I had my nose pressed against the outer window on one side to look at a model during an experiment, both opposite windows disintegrated, and flying pieces of heavy plate glass cracked the window on my side. Miraculously, my face was not touched.

After choosing a model, the operator had to decide on the Mach number at which a test was to be performed. During a given blast only one nozzle producing a single Mach number could be used, and so one selected a particular nozzle from the existing set of about half a dozen covering Mach numbers from slightly above 1 to 4.4, the highest available Mach number. The nozzle was put into the test section by an overhead crane with special prongs. The nozzles were boxlike units with equal external dimensions. Each nozzle's top and bottom converged at a constant 40-centimeter width from a low-speed supply section to a throat—the slitlike minimum area of the nozzle—at which sonic speed was attained. From there the top and bottom walls expanded again at a constant width to the 40-centimeter exit height, the location at which the exact test Mach number was achieved from wall to wall in the flow. The nozzles were carefully designed according to the known theory of supersonic flow. Their curved walls were made of a special plastic embedded between two brass templates, with the latter machined to close tolerances in order to provide a uniform flow (constant airspeed, pressure, and temperature over the entire testing area).

Finally, an optical apparatus was wheeled into place on rails mounted on top of the wind tunnel. Parallel light of high intensity produced by a high-pressure mercury arc (just like in some highway lights) and directed to large parabolically ground mirrors passed across the entire test system. The substantial temperature changes attending the flow about a model at such high speeds caused the index of refraction of the air to change. In turn, this variation was made visible by the deflection of the light, which was recorded as light or dark areas on a ground-glass screen or a photographic plate.[3] The details of the flow could then be studied. Pictures of the formation of the flow, its steady period, and its collapse still are impressive.

After adjustments in both tunnel and optical apparatus had been made, the operator pushed a button that activated the valve, and with a shattering noise a blast of air moved at supersonic speed through the test section. Pushing a second button terminated the run. One tried to perform experiments in as short a time as possible in order to keep down the evacuation time of the vacuum sphere.

A staff of about two hundred worked at the Aerodynamics Institute at the time of my arrival. Mechanical, optical, and electrical engineers, aerodynamicists, and technicians in many disciplines were required to run the operation. The institute was divided into several departments, and the aerodynamics department, led by Hermann Kurzweg, was the heart of the operation. Here the final product—the aerodynamic shape of missiles, artillery shells, and the like—was determined. In turn, Kurzweg, an exceptionally pleasant person, was in charge of a number of groups. Lehnert, who had introduced me to the place, directed the aerodynamics testing. The basic research group, of which I was a member, was directed by my boss, Erdmann. His work prior to my arrival consisted mainly of studying how to perform experiments at supersonic speeds. Another group, under Willy Heybey—an outstanding applied mathematician who worked in gasdynamics—provided the theoretical underpinnings. For example, the curved shape of the nozzles required to produce flow at several supersonic Mach numbers was devised by this group. Of the theorists in the group, I remember Wolfgang Zettler-Seidel, a young mathematician with a strong interest in psychiatry, a field that was in disrepute in Germany. Another mathematician, Elsbeth Hermann, was the sister of Günther Hermann, the assistant to the director. More on Günther later. Groups in thermodynamics, headed by Werner Kraus, and in optics, run by the husband-and-wife team of Ernst-Hans and Eva Winkler, were also attached to Kurzweg's department. In an outstanding machine shop run by Edmund Stollenwerk, any conceivable wind-tunnel model or other equipment could be designed and built. The aerodynamicist provided only the contours and a few additional instructions, and the engineers and the shop had to come up with an actual model or instrument.

The design and planning departments, which were parallel to Kurzweg's department and with which I had few contacts, completed the organization. I recall Heinrich Ramm, who was responsible for much of the instrumentation, especially the electrical and electronic aspects of this art. I also got to know Gerd Eber well when we roomed together in the United States. He worked on long-range planning.

The aerodynamicists Kurzweg and Lehnert had both received doctoral degrees at the University of Leipzig under L. Schiller, a well-known innovator in fluid mechanics. I believe both worked on the problem of the air drag of a sphere at low speeds. They jumped to supersonic flows as early collaborators of Hermann at Peenemünde. Erdmann had studied at what is now the Technical University of Berlin, and he joined the wind tunnels a little later than the others. In retrospect, I find it remarkable that this varied group, disregarding individuals' particular ranks in the hierarchy of the institute, worked together so smoothly. I never heard a harsh word: everyone helped everyone else, and good humor reigned; in fact, it was a pleasure to work in this place. Few breakdowns occurred, a fact that overshadowed the novelty of what was being done and gave me the impression of having joined a perfectly routine operation. Of course, I had never worked in a laboratory other than a teaching lab at the university. With this background, I was amazed that I was being let loose on certain experiments by myself rather soon after receiving what could only be called rudimentary instruction.

And so I settled into a new life in the office I shared with Pascual Jordan.[4] While I struggled with Prandtl's book and papers on the wind tunnels, Jordan sat in front of his typewriter all day composing a textbook on algebra without ever consulting notes. Usually toward the end of the day he would suddenly recall the original purpose of his assignment at Peenemünde. He had never worked in fluid dynamics, but rather than read books as I did, he proceeded to derive the equations of motion of supersonic flow. He delighted in the discovery of such phenomena as shock waves—well known in the literature but new to him—and I was awed by his ability. We often talked about other things, such as the history of physics in the twenties. He spoke in particular about quantum mechanics—his own major field, in which he had made important contributions. In addition, he talked about the politics of the twenties and thirties and the participants in the ongoing intellectual revolution, all of whom he knew well. He felt at ease criticizing the current ideology, and he explained to me its disastrous impact on the physics of our day.

I was of course the one who mostly listened, and we developed a pleasant relationship. Jordan had a serious speech impediment that was not so much a stutter as an inability to bring forth continuous sentences. I soon discovered that one had to turn one's back to him while he sat in his chair, feet on his desk and one hand on his forehead, twisting his nose with his thumb. These were the moments when he spoke rather fluently, and I was indeed fortunate to have been thrown into his company.

His actual assignment concerned a mixture of optics and fluid mechanics. An antiaircraft missile called by the code name *Wasserfall* (waterfall), which was supposed to fly at three times the speed of sound, was in the planning stage, including development of some components. Surely this was the weapon that my colonel at the Air Ministry had in mind when he called flak old stuff, and I will have much more to say about it. The scheme that Jordan studied was to embed an optical fuse in the glass nose of this missile in the hope that the infrared radiation from the exhaust of an aircraft could be pinpointed, followed, and used to activate the missile's explosive charge. A question had arisen concerning whether the radiation could pass through the boundary layer on the nose cone.[5] In this turbulent zone and at the high temperatures reached on the outer shell of a missile moving at supersonic speed, it was not certain that the rays of red light from the hot engine exhaust could penetrate to the glass wall.

At the time of my arrival in Peenemünde, several successful flights of the A4 had already taken place, and relatively few aerodynamic problems remained with this missile. One of these problems with which I became slightly involved concerned follow-up tests of the moving internal control surfaces in the rocket exhaust jet. These graphite vanes eroded quickly in the hot flames of the exhaust jet. An appreciable change in their shape occurred, and therefore their effectiveness as control surfaces varied along the rocket's trajectory. All of this was known from full-scale tests of the stationary rocket mounted in the test stands. In our wind-tunnel experiments, we used miniature brass models of the vanes during different stages of erosion. The hot exhaust was simulated by a high-pressure air jet released inside the model while the supersonic flow rushed about the outside. It turned out that one could, in fact, determine the varying aerodynamic turning moments exerted by these control surfaces. This and other minor tests provided the finishing aerodynamic touches to the A4. Kurzweg, who had been instrumental in the development of the aerodynamics of the A4, also supervised these experiments.[6] He was an imaginative individual from whom I learned much, in particular how to concentrate on the central problem rather than get immediately lost in details.

Now attention had to be turned to the Wasserfall and improved long-range missiles. Shortly before my arrival, work on the aerodynamics of the antiaircraft rocket had been assigned to Erdmann's group. The groundwork on the successors of the A4—the A4b and A10—and other projected missiles of varying designations was to be performed by Lehnert and his group. It was hoped, for example, to extend the range of the A4 dramatically by providing wings and speeding it up. The ambitious plans even included designs of a two-stage missile that would be able to cross the Atlantic. Erdmann was pleased with the distribution of tasks between the two groups in view of the obvious urgency of finding a defense against the increasing severity of Allied air raids. Surely producing antiaircraft weapons was a more realistic goal than the development of an intercontinental missile. Moreover, the antiaircraft missile was of special interest to the air force, which had no facilities to develop the aerodynamics of such a weapon, as seen from the previous listing of available supersonic wind tunnels.

Erdmann told me recently that after receiving this new job he had urgently requested additional help. Apparently this appeal was a factor in my assignment to the wind tunnels. Hans-Ulrich Eckert had also been brought in from a similar army unit at Peenemünde. The two of us were made subgroup leaders, designations that had no noticeable effect on our daily work. Hans-Ulrich became my closest friend at the wind tunnels, and our relationship continued during our joint work in Bavaria after the relocation of the Aerodynamics Institute.

The first major problem in conjunction with the Wasserfall was the aerodynamic control of the missile. The supersonic Wasserfall could of course readily outfly any World War II bomber or fighter airplane, all of which flew much slower than the speed of sound. (Even such famed fighter planes as the Spitfire and Messerschmidt flew substantially slower than a modern airliner.) Pursuit required quick turns to follow the zigzag of the slower aircraft, however, and such maneuvers presented serious aerodynamic and structural problems. Larger aerodynamic surfaces and larger rudders were needed, and in order to provide stability under a wide range of flight conditions the rocket was equipped—I believe at the suggestion of Kurzweg— with four additional pairs of stubby wings about halfway along its length. This must have been the first use of a shape commonly seen today in a variety of missiles.

The efforts of our group were also directed toward improving the controllability of this fancy missile by devising a variety of rudders. Most of this work, in contrast to similar studies today, was done by a combination of hunches about the design and wind-tunnel experiments on a purely empirical basis. No numerical solutions of the flow equations by computers were available—there were no computers other than electromechanical desktop calculators. The trusty slide rule still dominated the desk of the engineer, while the calculators were run by some of the many young female technicians at Peenemünde. The actual wind-tunnel models of the missile, usually only about a foot long, were masterpieces of mechanical craftsmanship. Since we could dream up different shapes of control surfaces more quickly than they could be made, the machine shop had to work hard in order to keep up with our stream of suggestions.

The second major series of aerodynamics tests on the Wasserfall concerned the distribution of pressure on the entire missile. From such experiments one could calculate the forces acting in flight on the external structure. These tests were started on November 24, 1943—long after the British air raid that will be discussed in chapter 5. Pressure is measured in such cases by drilling a little hole in the surface of the model, connecting it on the inside to a thin flexible tube, and leading the tube to a measuring device outside the wind tunnel. It turns out that it is impossible to cram a large number of little tubes inside a model about one foot in length and roughly an inch in diameter. In conjunction with the first experiments of this type on the A4, I believe it was Erdmann who had suggested slicing the model lengthwise and mounting it on a flat plate oriented in the flow direction. In this fashion, the many pressure tubes could be led outside the test section. Prior experiments had proved that the pressures measured at a given location on the so-called half-model agreed with those obtained from the real thing. Between designing rudders and determining forces and stability of the Wasserfall, we had our hands full.

In conjunction with the Wasserfall, a rocket that never even came close to completion during the war, there was one moment when I had a chance to enter military history, albeit in a truly minor capacity. A visit by Field

Marshal Wilhelm Keitel, the right-hand man of Hitler in running the Wehr-macht, was announced.[7] Models, graphs, and photographs were set up. The show was not unlike similar exhibits today in government and industrial laboratories, when funding agencies or the military send their representa-tives. Some of us were warned to be prepared to present our results, and one morning Keitel in fact appeared, surrounded by his staff. Somebody from the Air Ministry whose name I don't remember explained the Wasserfall to Keitel. Making hand gestures like those that fighter pilots use to demon-strate their pursuit of another airplane, he produced a vivid impression of the maneuverability of the Wasserfall, boasting that it would unfailingly search out its target and blast the enemy aircraft out of the sky.

At this point I entered the discussion. It seemed to me that we were discussing a world far removed from reality. I cautiously suggested that the anticipated success of the Wasserfall would depend on a long period of research and development, and that all aspects of the project were still in their earliest phases. Keitel listened to this with a serious face. Of course I never found out whether my remarks impressed him. It seemed to me that there was a general buildup of expectations from new, superior technology late in the war in the face of the deteriorating military situation. Again I must mention that this visit took place about five months after the disas-trous events at Stalingrad, which marked the beginning of the rapid down-turn of the Wehrmacht's fortunes.

Hopes and dreams of so-called wonder weapons were fostered even later in the war by the Propaganda Ministry; they were not shared by any of the technical workers I knew. But it seemed to me from the few contacts I had with high-ranking officers during this time that they got caught up in their own use of propaganda and consequently lost their sense of reality. Much of their effort and most of their time was devoted not to development or to the war per se but to efforts to secure funds, manpower, and equipment from the government for the great variety of new projects. Weapons—aircraft, tanks, and missiles—were being developed at a relatively late stage of the war. All had to compete for the rapidly shrinking supplies of raw materials and the decreasing industrial and labor capacity of Germany and the coun-tries occupied by it. Hitler's dream of winning the war with existing tech-nology evaporated, and in its place a multitude of often-wild projects arose.

At the Aerodynamics Institute, as elsewhere, many projects were se-riously initiated that did not have any chance of rapid completion. In con-junction with the plans for long-range missiles, such as the winged A10, a need for very high testing speeds arose. As I have noted, the current wind tunnels had their limit—remarkably high and unique as it was—at M=4.4. Before my arrival the staff for designing facilities under Eber had set to work on a new wind tunnel. An ultimate dream of supersonic aerodynamicists—a rare breed indeed back then—was to be realized. A supersonic wind tunnel with an enormous cross section by the standards of the time and capable of operating at very high Mach numbers was designed. The new tunnel was to run at first at M=7, and later the airspeed was to rise to M=10. It would be the first hypersonic tunnel in the world.[8]

The power to operate such a large tunnel was not available at Peenemünde, so those in charge planned to build the wind tunnel at the Walchensee Kraftwerk (power plant) in Kochel, Bavaria, which had begun operating in 1925. This hydroelectric plant at the foot of the Alps, the largest of its kind in Germany, made use of an altitude difference of 202 meters between two lakes, the Walchensee—the upper lake, at about 800 meters—and the Kochelsee. High-pressure water turbines were attached to a series of pipes between the lakes. The plan was to enlarge the plant substantially beyond its current power production of about 120 megawatts. In order not to deplete the upper lake with the additional power use, a tunnel was to be dug through the mountains to connect another lake in Bavaria and neighboring Austria. (This water project was completed after the war for the production of electricity for general use.) Several of the upper-level wind-tunnel scientists and engineers traveled frequently between Kochel and Peenemünde to make arrangements for building the new wind tunnel. These arrangements paid off handsomely, as we shall see later, when the need arose to move the existing Peenemünde wind tunnels and their shops, laboratories, and personnel to this relatively safe location in Bavaria.

There is little point here in relating more about the technical work done in the wind tunnels, but some anecdotes are of interest. Ludwig Prandtl visited the wind tunnels shortly before my time. I was told that he sat down in a chair in front of the ground-glass screen of the optical system while a model was installed in the test section. He pushed the start button, observed the unfolding flow pattern, and stopped the flow. He did this over and over for a long period of time. Although he had anticipated the panorama of such a flow many years before its realization, and although he had seen snatches of it in the tiny supersonic wind tunnel at Göttingen whose plan was copied for the Peenemünde facility, he was enthralled by the fine detail and the scale of the patterns that he could now observe.

Hermann Oberth, the intellectual father of rocketry in Germany, had in July 1941 joined the Peenemünde establishment at the invitation of von Braun, and he worked on a consulting and honorary basis. It was quickly apparent to everybody who met him that his great qualities were those of the inventor and the dreamer, yet his pioneering thoughts were rooted in a thorough knowledge of mechanics and those other aspects of physics and chemistry that were the essential ingredients of rocketry at the time. During his stay at Peenemünde, Oberth participated in the work of several technical areas. It appears that he spent much of 1943 at the wind tunnels, and I was associated with him for a number of shifts. He was a quiet, shy man, full of questions, and he showed a childlike curiosity about details that he was not familiar with. He insisted on helping during experiments by reading manometers and gauges, and if one kept pumping him, he slowly opened up and told interesting stories about his early thoughts. But more of him and the early history of rocketry in the next chapter.

My association with this new venture in science and technology—in which new methods developed with incredible speed thanks to the broad supply of high-quality manpower in all areas related to supersonic aerodynamics—was a fascinating experience. That experience awakened my interest in supersonic and hypersonic flows, which later widened to include applications of such flows to solve fundamental problems in chemical physics.

FOUR

History, People, and Special Events

T

HE VISION OF SPACE TRAVEL WAS THE FOUN-
dation of Peenemünde's past. That vision was held by a few
enthusiastic individuals among the leading members of the laboratory, and their dream of going to the moon led to the rapid
development of a large long-distance missile, the A-4.

Three rocket pioneers dominated the early work. Their thoughts—as far
as I know—were focused on travel outside the earth's atmosphere, artificial
satellites, upper-atmospheric research, and the like, not on the use of rockets
as weapons of war. It is important to remember that the Chinese fired
rockets in the thirteenth century, that men like Conrad Haas described
multistage rockets in the sixteenth century, that Walter Hohmann published
"modern" trajectories for moon flights in more recent times, and that others
in many countries have contributed. However, it is still fair to call these
pioneers the primary innovators of the twentieth century.[1]

The Russian Konstantin Eduardovich Tsiolkovsky (1857–1935) made
many contributions to aeronautics and to the foundations of space flight. By
the age of sixteen he had suggested the topic of interplanetary travel and
developed a theory of motion for a rocket that consumes its fuel and consequently changes its weight in flight. By 1900 he had designed airships,
monoplanes with wheeled undercarriages, efficient wings, automatic pilots
based on gyroscopes, and many other aeronautical objects. He also had
received a grant from the Academy of Science to build the first wind tunnel
in Russia. Early in the century he devised visionary plans for a liquid-
propellant rocket whose fuel is used to cool the combustion chamber. He
summarized his thoughts in a book, *Exploration of Outer Space with Reac-*

34

tive Devices, published in 1903. In its supplements Tsiolkovsky predicted artificial satellites, space stations, and other vehicles for flight beyond the earth. Despite his solid accomplishments—and probably because of his autodidactic past—his far-reaching visions were not accepted by the scientific establishment, where he was regarded as a dreamer. He was not taken seriously until after the Russian Revolution of 1917, when he was made a member of the academy and started to work on multistage rockets, jet airplanes, and air-cushion vehicles. He finally became well known in the 1920s, and his international reputation rose. Hermann Oberth admired his contributions and wrote to him. When space travel became a reality, a lunar crater was named after him.

Robert Hutchings Goddard (1882–1945) was born in Worcester, Massachusetts. He graduated in 1908 from the Worcester Polytechnic Institute, where he had begun to think about reaction engines, space flight, and aeronautics. He remained in Worcester to study physics at Clark University, where he received a Ph.D. in 1911. In 1914 he joined the physics department at Clark, and in 1919 he became a professor of physics. Goddard was among the first to develop the theory of rocket motion in a vacuum, much like Tsiolkovsky, who was not known to him at that time. But he also built practical devices, such as a rocket engine operated with liquid fuel, and he received many patents for his inventions. They included smokeless powder rockets, steering devices, multistage rockets, and many other items related to rocketry. A modest grant from the Smithsonian Institution led in 1926 to a successful firing near Worcester of the first liquid-propellant rocket in the world. This feat alone puts him in the first rank of the pioneers opening the path to space flight. Later, Charles Lindbergh took an interest in Goddard, supporting his work and helping him to receive funding from the Guggenheim and Carnegie foundations. Goddard moved his experiments to Roswell, New Mexico, where he fired a liquid-fuel rocket to an altitude of 7,000 feet (2,100 meters) at a speed of about 700 mph (1,100 km/hr). The firing was a major technological accomplishment. Goddard's work was not well known during his time because he was unwilling to share his thoughts with others. His reclusiveness kept him from being a team worker, an essential ingredient of progress in such difficult new ventures. In World War II, Goddard worked on jet-assisted takeoff of airplanes, research that also was under way at the newly founded Jet Propulsion Laboratory in Pasadena. Theodore von Kármán, the leader of the JPL group, attempted to share ideas on the subject, but Goddard rebuffed him. Goddard must nonetheless be counted among the triumvirate of rocketry pioneers, even if he had accomplished nothing but the successful firing of 1926. Goddard's importance was posthumously acknowledged when the National Aeronautics and Space Administration decided to name its major new postwar laboratory in Maryland the Goddard Space Flight Center.

Of the three pioneers discussed here, Hermann Oberth (1894–1989) exerted by far the greatest influence on German rocket enthusiasts. The young Wernher von Braun, who encountered Oberth in 1930 in Berlin, wrote, "I myself see in him the leading star [*Leitstern*] of my life, and in addition my first contact with the practical and theoretical aspects of rocket technology and space travel."

Oberth was born in Hermannstadt in Romania. His father, a physician, was asked two years after Hermann's birth to become the director of a hospital in Schässburg, a town of about eleven thousand in Transylvania, a mountainous region of Romania that had been settled in the twelfth and thirteenth centuries by Germans. Hermann and his younger brother grew up in a home where both father and mother fostered the boys' interests. For his tenth birthday Hermann received a small telescope and began to study the heavens. He asked about the moon and wanted to know how one got there. In high school he read Jules Verne's 1865 science fiction novel *De la terre à la lune* (From the earth to the moon), in which a moon ship is fired into space with an enormous cannon. Oberth calculated the implications of this scenario and proved that the crew could not survive the acceleration required to propel the ship beyond the gravitational field of the earth, a speed equal to 11.3 kilometers (7 miles) per second (25,000 mph). He decided that only a rocket could provide sustained acceleration at values that humans could tolerate. Calculations of this and related problems occupied him in school and at home, and he developed the theoretical framework needed to seriously consider space flight. His mother worried that he might fail the final school exams, the *Matura*, where knowledge in many fields outside the physical sciences and mathematics was required. But to her delight he passed his finals in 1912 with flying colors.

After leaving school Oberth began the study of medicine in Munich. His life, like that of many young men in Europe, was interrupted by the outbreak of World War I. He was drafted in 1914 and sent to the eastern front. In 1915 he was wounded and transferred to the military hospital in his hometown of Schässburg, where he immediately resumed his rocket studies. In 1917 he completed the world's first design of a long-range missile. A mixture of alcohol and liquid oxygen was to propel the rocket for a distance of 300 kilometers (190 miles). This project bears an uncanny resemblance to the final design of the A4.

In 1918, Oberth married, and one year later he resumed his university studies, now of physics. After studying in Klausenburg and Munich he arrived in 1920 in Göttingen, where he took courses with the outstanding mathematicians and physicists of this world-renowned university. Among them was Ludwig Prandtl, from whom he learned aerodynamics. Next he moved to Heidelberg, and though frequent switching of universities was common in his time, his mobility appears to have been rather high.

Oberth's proposed dissertation, "Die Rakete zu den Planetenräumen" (The rocket to interplanetary space), was turned down. It was considered too "fantastic" and "unorthodox." As noted by Barth, it was thought that the dissertation combined work done in too many different specialities, a now-recognized requirement in the field of rocketry. The famous Max Wolf, who taught astronomy at Heidelberg, said, "The dissertation is thoughtful and scientifically correct; however, sadly it is not really astronomy." The Nobel Prize–winner Philipp Lenard—who under Hitler invented the term "German physics"—found the thesis an outstanding achievement but not classical physics. The underlying skepticism about rockets and space flight pursued Oberth and others for years even as they were seriously engaged in the field. Unfortunately, people of dubious reputation outside science, including cranks, flocked to the subject. Even Oberth ran into trouble in trusting shady characters, as we shall see.

Meanwhile, the University of Klausenburg in Transylvania, where Oberth had started his studies, accepted him again. He submitted the theoretical part of his dissertation as a *Diplomarbeit,* and in May 1923 he passed his final exams. He had now fulfilled the educational requirements to become a teacher at a *Gymnasium* (high school). As was customary, he also received the title of professor. Late in 1923, Oberth succeeded in having his dissertation published by R. Oldenbourg, a publishing house in Munich. The author had to foot the bill himself, however. Fortunately, the book was immediately regarded as the seminal publication in the field. It was reprinted several times and became the standard reference work for the young generation of rocket devotees.

Oberth resumed his work, corresponded with Tsiolkovsky, kept writing scientific papers, and started experiments with liquid-fuel rocket motors. Promised funding by a German banker was blocked by a professor in Berlin, who like others before and after him did not even understand that rocket propulsion is based on Newton's law of "action equals reaction." This law of mechanics explains how rockets work in a vacuum by spewing out a high-speed jet of hot gases that provides a thrust in the opposite direction. On the personal side, a daughter was born in 1924, to be followed by three additional children. Fortunately, a position as teacher of physics and mathematics was offered by the Gymnasium in Mediasch in Transylvania, and the family moved there.

In the ensuing years, Oberth produced many papers and a new book on space flight. He began to mingle with the popularizers and ambiguous followers of rocketry. In 1929 he received the first international prize in astronautics from the French Society of Astronomy. A badly needed honorarium was included. Oberth also became the head of one of the burgeoning space societies and clubs. Fritz Lang's movie *Frau im Mond,* for which Oberth acted as consultant, was launched with fanfare in Berlin, and he participated in many other serious and not so serious undertakings. In 1930 the first testing ground for rockets was founded in Berlin, and there Oberth met von Braun. All in all, however, the 1920s and 1930s were difficult for him, his challenges ranging from the impact of the extreme inflation to the fight for acceptance of his ideas. Finally, in 1938 he received a stipend to perform experiments at the University of Dresden. In 1941 von Braun invited him to come to Peenemünde as a consultant; he stayed until the end of 1943, when I encountered him. After leaving the Baltic, Oberth produced a 1,300-page manuscript on all aspects of his field, a manuscript that got lost in the chaos of the war.

Immediately after the war Oberth was briefly detained by the Americans. On his release he joined his family in the town of Feucht, where he worked as a gardener until 1948. From 1955 to 1959 he worked with the Peenemünde group in Huntsville, Alabama, at the invitation of von Braun, who headed the development of real space rockets for the United States. Oberth's life became a series of visits, speeches, and events where he received honors. Other than the French prize, all of his thirty-three awards were bestowed after the war. They included honorary degrees from universities in several countries and the highest German civilian award, given to him by the president of the Federal Republic. Active to his last day, Oberth died at the age of ninety-five in 1989. There are now more than twenty biographical works on Oberth, demonstrating that he was an innovator of space flight who found a broad scientific and popular response.

Oberth's 1923 book was followed in Germany by an explosion of publications on rockets and space flight. The topic became the subject of serious and popular books and began to dominate many aspects of popular culture, including art and the movies. The idea of space travel took hold of the Germans. Michael J. Neufeld, writing in 1990 about what he terms the space fad, notes that in the 1920s the Germans were slowly recovering from defeat in World War I. A strong belief in the superiority of German technology and inventiveness arose, and it now included space flight, which had hitherto been regarded as utopian and bizarre.[2] Frank H. Winter, writing about the early rocket societies in all countries, collected a wonderful series of illustrations and documents showing the predominant role of German-speaking countries in the new field, including their influence in the formal organization of societies and clubs.[3] In the Soviet Union, Tsiolkovsky's followers were active, but the reticence of the Soviets in such matters precluded an open exchange of ideas with Western groups.

Activities that deal with the future have special appeal to the young. Emotions similar to those raised by the space fad were attached to all kinds of flight, including that of airplanes. I remember that I was touched by this effect when, with fellow students from my boarding school, I traveled to the Wasserkuppe, a mountain that had become the center of glider flying. This activity—more down-to-earth than rocketry—had inspired many other young people. The strictures of the Treaty of Versailles had severely limited motorized flight, but on this mountain we could view the birdlike soaring of wonderfully light sailplanes, including many of revolutionary design that had no tails or looked like the Concorde. Back at school we spent many hours building a functioning glider. But while the designers and pilots of the flying machines that we saw were serious, the rocket craze also produced strange practitioners of the cult, as we shall see.

The Verein für Raumschiffahrt (Society for Space Travel) was founded in July 1927 in Breslau (now Wroclaw, Poland). Aside from Oberth and later Wernher von Braun, Max Valier, an Austrian who liked rocket stunts, joined the club. Valier enlisted Fritz von Opel, the playboy son of the Opel automobile family. The two built rocket cars, and in 1928, before two thousand invited guests, they drove them on the AVUS, Berlin's famous automobile racetrack. Fritz von Opel also flew a rocket-propelled sailplane in public. The first showing of *Frau im Mond* had excited Berlin. Other related films followed, written by Lang's wife, Thea von Harbou, who was a well-known author. Neufeld recounts a typical event of the time that involved Oberth. As a consultant to Lang's movie company, he attracted another early space club member, Rudolf Nebel, who had been present at rocket tests, and a fugitive Bolshevik aviation writer, Alexander Scherchevsky. These three planned to construct a rocket to be fired off as part of a movie, but they vastly underrated the effort required, and the rocket never flew. Oberth called the Russian the second laziest man he had ever met, leaving open the identity of number one.

The excitement extended to children, who played with simulated rocket cars and airplanes. I still have a diary in which I sketched rockets and magical airplanes at the age of ten or so. Later, at school, I built real rocket vehicles. They consisted of one- to two-foot boards with four wheels, on which I mounted three homemade rockets—steel pipes filled with black powder, whose ingredients were purchased in separate stores so as not to arouse suspicion. The cars were equipped with fuses and let loose in a long basement corridor of one of the school buildings. At some point my helpers and I were stopped from further experimentation because great clouds of acrid smoke arose. Strangely, it never occurred to me or the teachers that such play was dangerous. My interests at school, however, soon turned to geophysics and astronomy, and the rocket cars were my last contact with such activities until I joined the Peenemünde establishment.

Nebel had built a "rocket port" near Berlin. His group included von Braun, Klaus Riedel, and Helmut Gröttrup. The last two had also been members of the space society, and later von Braun invited them to join him at Peenemünde, where they played a major role. In 1929 popular enthusiasm petered out. Valier died in an accident involving a liquid-fuel engine, a fact that led the authorities to control the experiments. The devoted believers in space flight, however, including those who later played an active role in Peenemünde, did not give up their research. In spite of the disjointedness of experimentation at various locations and under various auspices, including Dornberger's army effort, serious technical headway was made. The stage was set for the future, and the prior work sped up the later A4 development. Ironically, some time was lost when Hitler stopped rocket experimentation in 1933.[4]

When I started there in June 1943 I knew very little of what I have just described, plunging into the new environment with no preparation. I feel some trepidation in describing associations and events, not least of all because it is difficult indeed to separate actual recollections of times long past from later readings and stories heard afterward.

Life in Haus 30 turned out to be interesting and comfortable. I was billeted there presumably because I was an officer, and I thus got to know the unmarried civilians who led many of the departments. Klaus Scheufelen, another officer from an antiaircraft unit, lived there. Like me, he worked on the Wasserfall, but his job was the development of the rocket motor for the missile. This task may have been more exciting and strenuous, yet I preferred my quieter aerodynamics activities. My accommodations allowed me to meet some of the staff—including von Braun—outside the narrow circle of the wind-tunnel people. I lived only a short time in Haus 30 because after my marriage in August I moved to Koserow (see chap. 5), but it was long enough to put me in touch with the center of activities, a connection that would last to the end of my stay in Peenemünde.

A small group of friends who had known each other for some time graciously accepted me as a member, and we talked during many evenings. Eberhard Rees, the administrative director of the place, came from Swabia, in the south of Germany. He displayed the liberalism and restrained humor of that region of Germany, which manifested itself in a seemingly unlimited repertoire of funny stories, often involving the volunteer fire department of a small village. An older army captain was the *Abwehroffizier,* or counter-intelligence officer, who had to watch out for spies. In civilian life he was a judge in Pomerania. His strict sense of decency had made him severely critical of the destruction of the legal system by the government. We often joined him to listen to the BBC news on the radio. He had official permission to hear the enemy's version of events and consequently possessed the proper radio equipment. It was characteristic that none of us ever considered that someone might report activities like this to the authorities. I met Ernst Steinhoff, who was responsible for the guidance system of the A4, and I later encountered him in the United States. Steinhoff's brother, one of the most successful pilots of the German air force, became the head of the new air force after the postwar rearmament of the Federal Republic. Most of the leading scientists were about ten years older than I. This was also true for those at the wind tunnels, excepting Erdmann, who was roughly my age. Von Braun, who headed the army's projects from the beginning, was only five years older than I.

None of the Haus 30 people that I saw privately believed in the *Endsieg,* the final victory. All saw the situation in Russia for what it was, expected an Allied invasion in the west, and discussed in small groups which army would be where at the end of the war, particularly the Russians. I do not want to suggest that most Germans, or even most of the people at the laboratory, thought about the war in these terms. In fact, I am sure that the majority of the people at Peenemünde did not think too much about the future, applying all their energies to the work at hand. But I am struck again in retrospect by the fact that none of us ever seriously considered the threat of an air raid on the laboratory, even though we heard the nightly droning of the bombers flying toward Berlin.

Most of the technical personnel were caught up in the task of completing the work on the A4 and later pushing the development of the Wasserfall. Practically all discussions in larger groups—in the mess hall and in other more public places—concerned technical problems. Everybody spoke freely of his work; internal secrecy simply did not exist. *Fachsimpeln,* or shoptalk, squeezed out all other discussions. Whatever personal opinions might have been held by individuals, the support of the war effort was uncritical: the technical work had to be completed in the shortest possible time.

During my time at the Baltic I never heard a single remark about space flight. This is puzzling, for several reasons. For one, some of the early rocket designers were in leading positions at Peenemünde. Even Oberth, the father of them all, moved around as a consultant. But no one ever mentioned in my presence that the A4 would be a steppingstone toward a moon flight. In my several meetings with von Braun, he never suggested this possibility, even in small social gatherings. In contrast, much of the postwar literature recounts the A4 development as if space flight had been a distant goal discussed during the war. A typical example is the book by Klee and Merk (note 4), who characteristically named a chapter "Der Umweg über die Waffe"—that is, the detour via the weapon. An interesting exception to my observation is suggested by Richard Lehnert, who writes that soon after his arrival at the Aerodynamics Institute in April 1937 he visited the library, which had just started to accumulate books.[5] (I have to confess that I never knew there was a library at the place.) To his surprise, he found volumes on rocket development and space flight. By chance, Dr. Wernher von Braun, as Lehnert calls him, entered. Lehnert, referring to the space flight material, thanked him for the foresight he had shown in adding recreational literature to the technical library. Von Braun was surprised that Lehnert took this view, since to him this was serious stuff. Then he laughed when Lehnert told him that Hermann, Lehnert's boss, had not said a thing about the purpose of the laboratory, keeping quiet presumably for reasons of secrecy. After further discussion, von Braun stated that the Peenemünde work was just the first step toward space flight. Lehnert further recounts that only half a year later, speaking in intervals between firings of forerunners of the A4 from the island of Oie, von Braun gave an informal talk to his associates about his completed plan for a manned moon rocket. To impart a feeling for the size of the required carrier rocket, he used a visible object: the tall lighthouse on the northern tip of the island. Lehnert concludes by recalling that it took only thirty-two years to verify von Braun's predictions. This story is no doubt true, but by the time I showed up, other considerations were obviously preoccupying the staff.

The large sums spent at Peenemünde, as well as the numbers of scientists and engineers who were tied down by the A4 development rather than working on a more conventional weapon, such as a large bomber, were never mentioned in discussions that I witnessed. Nobody ever voiced a doubt about the enterprise when I was around. Of course, at such times people do not divulge their innermost thoughts. After I understood what was really going on at Peenemünde, however, I began to wonder about the ability of the A4 to affect the outcome of the war. Every A4 carried about one ton of conventional explosives in its warhead. In addition, the impact of the rocket hitting the ground at supersonic speed added much to its destructive effect. Even in experiments without explosives, the missile produced a large crater. But how many of these complex machines could be built? Here I vastly underestimated the remaining German industrial capacity, as I know

now. Also I wondered how inexperienced artillerists could pour liquid oxygen into the tanks, adjust the gyroscopes, and so forth, so that the A4 would hit a distant target. More important, this was the summer when such major German cities as Hamburg were being bombarded by hundreds of bombers nearly every night. Each of these aircraft carried several times the explosive load of a single A4. Looking at this fact alone, I wondered whether similar thoughts had not occurred to the general staff, and whether suddenly Peenemünde would close down or switch to producing different weapons. Fortunately for me and many others, that never happened.

Coincidentally, the laboratory kept many healthy young—and not so young—men from being drafted into the Wehrmacht. Since the army continually suffered substantial casualties, especially in the east, great efforts had to be made to replenish the ranks. Moreover, soldiers like me working in science or industry could be reassigned to fighting units with the stroke of a pen. Again, I believe, little thought was given by anyone at Peenemünde to such scenarios, since civilians and soldiers alike had been assigned to jobs that turned out to be an engineer's dream. It would also be unreasonable to expect anyone—including me—to rebel against an assignment to produce novel weapons of war—or indeed any weapon—that would keep Hitler in power. However, I was still struck by the lack of any private discussions about the effectiveness of the A4 in relation to the war effort. Rumors of wonder weapons began to circulate much later, reaching a crescendo by the end of 1944 or so. The rumors may or may not have been spread by Goebbels, the minister of propaganda. It was said that if the Führer would give the signal, the war would quickly be brought to an end. Anyone who had worked at Peenemünde surely knew that the A4 was not the weapon that could do such a thing.

Many of the civilians who were not already members had joined the National Socialist—or Nazi—party in 1938, the year that a great wave of apparent conversions boosted membership. That was the year when pressure was exerted to increase party membership in the civil service, in the universities (run by the states in Germany) and other governmental or near-governmental organizations, in many industries, and even in the arts. Entry into the party more than five years after Hitler came to power should not necessarily be correlated with enthusiasm for the man and his government. This is true in spite of the stunning success of his foreign adventures. Many believed that joining was essential to retain their jobs; others who had been skeptical were in fact truly converted by 1938. For example, a few of the hesitant teachers in my boarding school became party members, and, I suppose, so it was with many at Peenemünde. Quite aside from convictions, opportunities opened up: ambitious people could make headway only as party members, and others were simply drifting along. This mixture of motives and the fact that many joined the party at a relatively late date made it difficult for the Allies to determine after the war who was actually a "Nazi"; certainly filling out a questionnaire would not do the job. I am sure that it is not easy for those who have not lived in a similar environment to understand the difficulty of categorizing people whose behavior is often ambivalent. At the wind tunnels I did not knowingly encounter a serious disciple of Hitler—with the exception of the director—among those few I dealt with daily. It appears that Rudolf Hermann had some party function at Peenemünde, but I am not sure about its exact nature. Later, after the wind tunnel had been moved to Kochel, I discovered who headed the party at the new location. At Peenemünde, however, I did not feel a politically oppressive atmosphere. Many of the staff were not party members, no ideological influence was ever applied, and the work took precedence over all other activities.

Aside from technical discussions about the daily work and its frustrations, I was occasionally exposed to ideas about planning for the future. These plans had no relation to space flight, being largely directed toward increasing the distance covered by the A4. These conversations were quite free. I am not sure whether any security clearance procedures were ever applied to us, but I must assume that some kind of check was made, because we dealt with documents stamped *Geheim,* or confidential, and even *Geheime Kommandosache,* something like top secret. In addition, documents had special numbers assigned to them that were separately recorded. My lack of experience prevented me from contributing anything of substance, and I simply enjoyed the uninhibited talk. Discussions crossed the departmental links of our work, and no "need to know" concepts of the sort that I later encountered in the United States applied.

Configurations of the planned long-distance missile were tested in the wind tunnel, and later in firing tests. Among their code names I find A4b, A9, A10 and A10b; of these, the A4b was an A4 to which large wings had been attached. This rocket was also called a glider, because the wings and larger aerodynamic control surfaces could increase the distance from firing location to the target. After a nearly vertical firing, a gradual reentry of the denser atmosphere, not unlike that of the current space shuttle, was envisaged. Another planned trajectory pursued a wavy path out of and into the atmosphere, like that of a flat rock skipped on a lake. Since it had been demonstrated that rockets could be safely fired under water to arise vertically from the surface, another scheme foreshadowed the current Trident submarines. A series of regular A4s was to be packed individually into watertight containers that would float just below the ocean surface. The containers would be pulled in a string by a submarine across the Atlantic. This convoy would stop near the coast of the United States, where the water ballast would be redistributed and the containers tilted upright. At that stage, the missiles would be adjusted and fired at New York City. I do not recall how anyone planned to fill the missile tanks with fuel and liquid oxygen at temperatures of about −200°C (−330°F) while they were floating in the waves. None of these or other schemes, such as the two-stage rocket already mentioned, ever entered a serious development phase, as far as I know.

At some date I was promoted to *Oberleutnant*, or first lieutenant, and I also received a second-class medal for civilian contributions to the war effort. The officer rank still had the great advantage of enabling me to participate in the courier service to Berlin, permitting occasional visits with family and friends. On one of these trips late in the year I experienced my first arrest during the war. On the way to Berlin the train often made a long stop in the small town of Pasewalk in Pomerania. I used to get out and admire the impressive waiting room built in the late nineteenth century, where I was particularly taken by an enormous bar with cases full of glittering glasses. (During a visit to the station in 1992, I found that the room had been largely demolished to house a pizzeria.) On this particular occasion I was approached on the platform by an older noncommissioned officer of whatever top rank one could achieve in the service. The uniforms of such individuals looked more impressive than those of generals. He assumed a military pose, identified himself as the commander of the station, and said that I was under arrest. I wondered what I had done. He demanded my papers—the *Soldbuch*, the identity booklet that all soldiers carried—and checked personal data. He next stated that I had been pointed out by someone who said I was a spy. I was forced to pull out the papers identifying me as a courier who had to be assisted by all authorities, and so I talked my way out of this embarrassing situation. Since he had mentioned a young *Flakhelfer*, the reason for my arrest was apparent to me.

Late one evening I was returning by train from a visit to the University of Greifswald. In the darkness imposed because of the blackout, I noted some youngsters in the compartment. Adolescents were drafted to operate antiaircraft guns, and my companions had been assigned to my old 20-mm guns. I engaged them in talk about how things were these days in the flak. Had they actually fired their guns in practice or even at airplanes, how was their training, and so forth. One of the kids had obviously seen me on the platform now, and his indoctrination had taught him that even his little cannon's workings had to be guarded against the likes of me. His was the generation that during the final phase of the war frequently fought to the last moment. The soldiers' short lives had been spent in devotion to Hitler, and I wondered how they would feel in a year or two. There exists a photograph in which a disheveled Hitler pins Iron Crosses on the too-large coats of children who fought in the Battle of Berlin; there are also photographs of such kids who broke down and cried in the face of the violence.

I remember with great pleasure the courier weekends in Berlin at my father's house. Invariably I encountered people who knew much of what went on behind the propaganda. Topics ranged from the many love affairs of Goebbels, who was responsible for the film industry, among many other functions, to the real status of the war. At one point my father received the title of *Staatsschauspieler,* roughly state actor, an honor that included a life-size sculpture of Goebbels's head. My father promptly placed this gold-painted work of art in the downstairs lavatory, and many a guest was startled on entering the room. I recall an afternoon of conversations with Theodor Heuss in the summer of 1943 in which he predicted the course of the war. I had not known him before, though his son Ludwig had been a friend of mine at boarding school. Heuss, a liberal Swabian, spent his time writing biographies. He had finished a book about Anton Dohrn, the founder of the marine biology station at Naples and a relative of ours. At the time of Heuss's visit, the Japanese had moved close to Australia. Heuss suggested that they were overextended and would have to pull back ships and troops. He predicted the Allied invasion for the middle of 1944 and outlined the further collapse of the Axis powers and the Third Reich in remarkable detail. But even Heuss did not foresee the possibility of an assassination attempt on Hitler by military officers, and he did not realize that Germany could hold out until May 1945. Everybody was at a loss when talk turned to ideas about the grim future of Germany after the war. Little could those in the room have anticipated that after the war Heuss would become the first president of West Germany, with Konrad Adenauer as the chancellor.

To turn to a less serious subject for a moment: one day I saw an automobile of startling design, with a rounded front and a rounded back. It was a convertible driven by Dr. Robert Ley, the head of the Arbeitsfront, a labor union in which workers had no vote. Strangely, Ley, like Goebbels, was always called doctor; that may be a reflection on the dearth of academically trained people among the men who surrounded Hitler in the early days. The contraption I saw was indeed the first Volkswagen Beetle. Of course, the standard Beetle was not a convertible, but after the war such a model was sold. Millions of people in Germany opened special accounts to save up for a VW; their money went directly into the war effort. Who could have foreseen that this car, baptized by Hitler as "Volkswagen," would capture the whole world and be produced in greater quantity than the mythic Model T Ford?

During the war, however, the chassis and engine of the VW were fitted with a jeeplike body and served in the armed forces.

During my stay at the Baltic—in particular during the summer of 1943—I saw much of Wernher von Braun. I occasionally encountered him at technical meetings to which I was sent to watch out for aerodynamics, and on informal social occasions. (I do not remember formal dinners or other festivities arranged to entertain the many guests who wanted to see a rocket firing. Most likely such affairs did take place, in view of the importance of impressing the leadership for the continued support of the laboratory. Young newcomers would obviously not have been invited.) Von Braun came from an old family in Silesia, Germany's eastern coal and steel region, now part of Poland. At the time of his birth, his father, Baron Magnus von Braun, was the head of a district government. Later the elder von Braun became a minister involved with agriculture in two German cabinets prior to Hitler's time. When Hitler came to power, he resigned from government service. Once I briefly met von Braun's parents, who were pleasant and self-effacing. His mother had musical talent. Wernher, her oldest son, played the cello, and with Heinrich Ramm (chapter 3) and others he participated in a chamber music group at Peenemünde. In spite of his ancestry, Wernher (a younger brother had his father's first name) never showed signs of having the slightest aristocratic snobbery, although such snobbery would be common in someone of his background then and now in Germany. He went to two boarding schools that were founded by Hermann Lietz, a major educational innovator. I mention this fact because my own boarding school was an offspring of a Lietz school opened in 1908. These coeducational schools were exceptionally liberal and free of restrictions—religious or otherwise—in contrast with the high schools run by the government. Von Braun developed a strong interest in astronomy, and after finishing high school he became a physics student at my alma mater, Friedrich-Wilhelm-Universität in Berlin, now called Humboldt University. In April 1934, at the remarkably young age of twenty-three, he received a doctoral degree for a dissertation on the theoretical and experimental problems of liquid-fuel rockets. Not unlike Goddard, he first thought of using rockets for high-altitude research. Soon his plans extended to space flight, however—plans that were strongly influenced by Oberth. All this was laid aside during the years at Peenemünde because he needed to concentrate on leading a huge laboratory founded to develop rockets as weapons of war. As is well known, his early dreams were eventually fulfilled by his work in the United States. He wrote much, and more has been written about him.[6] But in this narrative I will stick to my own experience of him.

I can attest to von Braun's leadership at Peenemünde. It was von Braun who assembled the core team needed for missile design and construction. He had a vision of the special talents needed for such a new enterprise. He himself was a good general physicist, and he had exceptional engineering talent, a rare combination. In 1936, the year he began directing the A4 project, the underlying science of this task was largely in hand based on Newton's laws in conjunction with more recently explored areas, such as trajectory calculations, aerodynamics, atmospheric properties, the chemistry of propulsion systems, electronics, and gyroscopic guidance. But putting this knowledge together, filling in such unknown areas as supersonic flow, and devising such previously unheard of apparatuses as a high-speed pump that could spray fuel and liquid oxygen at high pressure into a combustion chamber were indeed remarkable achievements. All this had to be done quickly, and it required directing a large staff.

At a meeting of people from several technical areas it quickly became apparent that von Braun knew more than anyone about the many ingredients of missile design. Every specialist sitting at the table was better versed in his own field, but von Braun had a remarkable grasp of all the fields. He could separate important from peripheral items, distinguish what had to come first, make clear decisions, and inspire people to work. He was close to being obsessed with rocket development. In small groups, however, he could relax and detach himself from rocketry. Once the conversation turned to Rome, a city that I had not yet visited. Von Braun remarked that the Coliseum ought to be demolished as a terrible impediment to the flow of traffic. I was not sure whether he really meant what he said or was joking. The term *technocrat* has come up in conjunction with von Braun, but I am certain that he cannot be dismissed by such typecasting.

I noted that some of von Braun's close associates had a tendency to flatter him, a fact that he himself did not seem to notice. After he received the title professor, which the German government can dispense even though a person is not associated with a university, he was called Herr Professor by his entourage. Despite holding a powerful position in a semimilitary environment, he remained remarkably relaxed, did not flaunt his authority, and got along well with his military bosses. Fortunately for the project, he had a close personal relationship with Dornberger, the army boss of the enterprise. Dornberger was by now a colonel, while another capable and pleasant colonel (later general), Leo Zanssen, a member of the early military group, actually ran the military side of the place. It was Dornberger, with von Braun by his side, who had persuaded Hitler at a critical juncture in the affairs of Peenemünde to keep up the A4 development. This was accomplished largely by showing films of successful launches. Von Braun joked in small groups about meetings with government leaders, and he extended this attitude later to the SS. It became obvious to me that he disliked Hitler and all that Hitler did. But it was Hitler who supported his dream. In contrast to Lehnert, I never heard von Braun speak about space flight, however. I have no idea what he thought about the military value of the missile: surely he of all people considered the dubious effectiveness of the A4.

I was still in Peenemünde in the spring of 1944 when we heard that von Braun had been arrested by the SS. I now know that Klaus Riedel (there were three Riedels, I believe) and Helmut Gröttrup—who, after the war, led the German rocket group that worked in the Soviet Union—were also taken into custody. Dornberger called Keitel, the top man at Wehrmacht headquarters in Berlin, who said that the matter was out of his hands. Apparently even the highest military men were afraid of Himmler and his tight organization, which looked more and more like a state within the state. But Dornberger went to Berlin and appealed to all the leading government officials he knew. He also went to the SS, a move that certainly was not without risk to him. He succeeded in freeing von Braun and his colleagues, and he even learned why they had been detained. The three were accused by the SS of not applying their total effort to the development of the A4 because they were working on preparations for space flight. Of course nothing could have been further from the truth, considering the reality of the work at Peenemünde. After all, even a casual visitor to the development station had to notice the enormous effort being expended to speed up production of the A4 in parallel with testing to eradicate the remaining technical problems. This effort was led by von Braun, who possessed seemingly unlimited energy. The arrest appears to have been part of the devious efforts of the SS to achieve control of the A4 and other projects involving new weapons, efforts that were ultimately successful. When the prisoners were freed to return to their work, the release was restricted to a three-month period. Strangely, at the end of this time nothing further happened.

A more pleasant memory leads me to relate a spy story that is most likely apocryphal. A woman dentist at Zinnowitz, the closest resort town to Peenemünde, kept an open house. Zinnowitz was on the Usedom rail line, and a number of us—including von Braun—were occasionally invited to her parties. I vaguely recall that one could also just show up at certain times. This attractive woman, as I remember her, had a large house and served plenty of food and drink at her gatherings. The parties were a pleasant diversion from work. I recall that one evening another young officer who was returning from a courier mission to Berlin joined the party before delivering his secret satchel to the authorities, as required. He drank too much and ended up asleep on a bed, and as a practical joke we removed his briefcase. When he woke up, he was suitably startled and appalled at his neglect of duty, but we did not keep him in suspense for long.

I later heard that some believed the dentist worked for the Soviets; clearly she could pick up a great deal of information from chance remarks at her parties, and she might even have had an opportunity to inspect a courier's satchel, which must have contained interesting data. Whatever the truth of this story, I never wondered at the time how it was possible to procure legally so much wine and food. There was a thriving black market in Berlin in spite of the severe sentences—including death—that threatened dealers. My father used to buy butter and other food at astronomical prices to augment the by now meager supplies obtained with ration cards. My lack of concern about the source of such lavish spreads is another instance of my naïve pragmatism, an attitude that undoubtedly made life easier. In addition, I was told that the dentist worked with a high-ranking civilian at Peenemünde in the employ of the Soviets, another spy rumor that I have left to the historians. When I thought later about this story, which made for interesting conversation at the time, a simple alternate explanation emerged. I would not be surprised if the perceptive lady had early on anticipated shortages of dental material, including gold. Later, rather than being paid for fillings, bridges, and the like, she bartered for food supplies. The hinterland of Usedom was an agricultural area that made this possible. German farmers never suffered from hunger; they even ran distilleries and lived the good life.

At a time when the firing of the A4 was becoming increasingly successful, a strange phenomenon appeared shortly before impact at the end of the trajectory. About 1,000 to 1,500 meters (roughly 3,000 to 5,000 feet) above the ground, the entire missile exploded! The test firings at Peenemünde were, of course, done without the explosive warhead, so the problem was not in that part of the missile. However, after the preset, exactly timed shutoff of the burning phase, some fuel and oxygen remained in the tanks and piping leading to the combustion chamber. By what mechanism was this material being ignited? It was understood that at the A4's maximum Mach number of above 4, the skin temperature of the missile exceeded 700°C (1,300°F) for a short time during the flight through the stratosphere. On the other hand, cooling of the skin by heat conduction or radiation, together with a varying external atmospheric state, made it difficult to estimate the temperatures inside and out. Did tanks burst under the influence of the heat, permitting fuel and oxygen to mix and explode? Did some other material ignite, owing to aerodynamic heating prior to the explosion? (The term aerodynamic heating relates to the warming of the missile's outer skin through air friction.) This and the other problems involved in calculating the surface temperature can now be solved. In fact, they have been solved for much more extreme conditions than those affecting the A4, namely the famous reentry of the space shuttle. But these advances took place many years after the period under discussion, and the cause of the A4 explosions was hotly debated. It was decided to perform two types of experiments to find a solution to this problem, and I participated in one of them.

The firing range at Peenemünde was too short to allow the positioning of observers and instruments at the end of a standard trajectory. Recall that the A4 firings extended over the Baltic, more or less parallel to the coast. As an alternative, it was decided to fire vertically upward and hope that the returning A4 would splash into the ocean close to the shore. A handful of people of different disciplines, including von Braun, Hermann, and me, were placed on a wooden tower with a platform on top. The A4 was fired straight up from its main testing arena, and it quickly disappeared into high clouds. An electronic Doppler signal—which indicated the relative speed of the missile away from us, and later toward us—was translated into a tone of changing loudness and pitch emitted from loudspeakers on our platform. The invisible missile consequently produced howling noises as it flew upward to the outer fringes of the atmosphere. *Brennschluss*, the cutoff of the rocket jet, was signaled, and the momentum achieved at that stage of flight kept the A4 coasting upward. The noise became quieter and quieter as the missile slowed, until it ceased altogether. A strange quiet also settled on the observers. We knew that gravitational attraction had made the missile stop at an altitude where the atmosphere was barely perceptible. Next the descent started as a free fall to earth, since practically no air drag existed up there. Slowly the noise from the speakers returned. As increasingly dense layers of air were encountered, the A4 began to tumble, and finally it turned over, with its nose pointing downward toward us. This behavior was a result of the wind-tunnel design that made the A4 aerodynamically stable, like a feathered arrow. The rocket picked up speed, the noise from the speakers increased, and the decreasing altitude was called out by a technician. I hoped that the flight would end in the water, close to the shore as planned.

We had formed a circle facing outward, and the first person to see the missile exploding at low altitude—provided that it indeed malfunctioned— was to shout for all of us to turn. Would we be able to see the disintegration? Would we observe sufficient details of the explosion to obtain a hint about the underlying mechanism? At any rate, the experiment scared me, though the chance that the tower would be hit was small. The explosion did in fact take place within the range of clear sight. I did not catch the initial blast but saw a cloud after rapidly turning around. I forget who the lucky primary observer was, but I recall that nothing much was learned from the experiment. Fortunately, the missile landed in the water and, I believe, did no damage. I heard later that some of the A4s fired at England were subject to the same low-altitude explosions, magnified by the explosive warhead. It was thought that a new feature of unknown purpose had been added to the missile design. This fascinating experience was the only one in which I had direct contact with the A4's development. Here I saw a demonstration of what was encountered by the early pioneers as well as by those doing engine testing and related work. In contrast, wind-tunnel experiments—though great fun—had an academic aspect of peaceful laboratory exploration.

A second set of experiments was performed in eastern Poland, though I only heard about them. The local German authorities were asked by the army to remove the residents of a village and its surrounding area in a relatively sparsely populated target zone. While the test missile was set up to fly its entire trajectory over populated stretches of land, it was aimed at the steeple of a church in the selected village. Since one instinctively did not believe that the A4 would exactly hit its target, von Braun and his associates took the unscientific step of locating their observation point in the steeple. Ernst Geissler, the cool-headed leader of the mathematics group in charge of trajectory calculations, pointed out to me that the church tower was still the most likely point to be hit. Several missiles were fired; some exploded before they hit the ground, and the explosions were observed from the steeple. Again nothing much was accomplished. Nobody got hurt in the process, and further work was done in the laboratories and by calculations. I believe Lehnert's aerodynamics group performed new experiments in the wind tunnel, even later at Kochel. Unfortunately, however, in supersonic wind tunnels of the Peenemünde type the temperature effects of real supersonic flight could not be simulated. Nevertheless, since many other factors may have contributed to the strange explosions, the work was useful. The decision to add additional heat insulation on the inside of the A4 resulted in a substantial improvement but not in a cure for the exploding rockets.

I saw von Braun once more in Germany shortly before the end of the war. Later I met him on several occasions when he visited Washington, D.C., where I first worked after the war. I also saw him at the Marshall Space Flight Center in Huntsville, Alabama, where I stood looking in awe at the gigantic Saturn V carrier rocket that eventually put men on the moon. On these occasions von Braun tried to interest me in joining his group. I resisted because my career goals and personal interests were incompatible with the aerodynamics of missiles and rocket development in general. Moreover, without any adverse judgment of the people involved, I wished to be independent of the group of German scientists and engineers that surrounded von Braun. My last meeting with von Braun was a chance encounter early in the morning at an airport, I believe in Dallas. He renewed his offer, and again I had to say no. I was saddened by the early death in 1977 of this gifted and complex man.

FIVE

The British Air Raid

FTER SETTLING IN AT PEENEMÜNDE, I FOUND
my personal circumstances safely stabilized for the first time in
several years. No immediate resumption of real soldiering seemed
likely, so it did not appear to be unreasonable to get married. My
future wife and I had met at boarding school, from which she
graduated one year after me, in 1937. After that she studied classical archae-
ology at the University of Freiburg in the south of Germany.

We decided to marry in early August, a decision that was not easy to
carry out. The city of Berlin, the planned site for this event, had suffered
heavy air raids toward the end of my second study leave, and other cities had
been attacked by large numbers of aircraft. In particular, in the summer of
1943 the port city of Hamburg experienced a devastating new phenomenon
occasioned by heavy bombing: a firestorm. Civil defense procedures that
had protected people were now of little use. Shelters in basements had
provided some protection, and water stored in buckets and bathtubs and
sand stored in boxes aided in fighting smaller fires. Relatively speaking, few
deaths occurred. However, the conflagration of a firestorm, which arose
when huge numbers of incendiary bombs were dropped after an urban area
was attacked by explosives, consumed much of the oxygen at street level,
with devastating consequences for the population confined to the shelters.

After the Hamburg experience became known, it seemed that for the
first time Berlin was gripped by something close to panic. This excitement
was in large part caused by Joseph Goebbels, the *Gauleiter*—the top Nazi
official—of Berlin. Goebbels ordered an evacuation of the city by all who
were not part of the work force. Entire schools, with their students, teach-

ers, and equipment, were packed up and sent to small towns all over Germany. There was a ban on nonessential travel to Berlin, and so it appeared foolhardy to try to assemble two extended families for a wedding. Despite these obstacles, all our relatives made the trip successfully, and the ceremony took place at the office of a registrar, followed by a sumptuous dinner. The Hotel Bristol, one of the fanciest in Berlin, had survived the air raids up to that time and could be bribed to provide the required food and wine for close to twenty people.

After the wedding, we went for a few days to Heidelberg, traveling on a crowded train. My wife climbed through a window to claim two seats while I fought through the aisle with our luggage. We found little peace in Heidelberg because of a series of air attacks on the nearby city of Mannheim. One could see the fires from Heidelberg, itself a picturesque town not touched by the war then or later. While we were standing outside on a hill one night, an 88-mm flak shell exploded close to us, the automatic fuse that was supposed to detonate the shell in the air having failed.

Before leaving Peenemünde for the wedding, I had moved out of Haus 30 and rented a room in the Hotel Deutsches Haus in the village of Koserow, one of the many vacation spots on Usedom, about thirteen miles from the laboratory on the train line that ran parallel to the coast. My wife matriculated at the University of Greifswald, situated in a beautiful small town that could easily be reached by a short train ride. During the week she stayed at the university and I commuted to Koserow, where she joined me for the weekend. This deceptively regular life turned out to be short-lived.

At that time, the wind tunnels operated in two shifts, with the night shift extending past midnight. I had by now acquired sufficient experience to be trusted occasionally with responsibility for the night shift in the absence of other aerodynamicists. When I worked the night shift, I had to stay overnight at the laboratory, since no trains operated that late. A pleasant room in a single-story frame building was provided for me, and it was there that I kept most of my things. In the morning after a night shift I left for Koserow and enjoyed the beach for the day, just as I had during my vacations as a boy.

On the evening of August 17, it was again my turn to work at night. I don't remember what experiments we planned to perform. In addition to a skeleton staff of technicians to run the pumps, operate the overhead crane that changed the nozzle in the test section, install models, and assist in case of technical malfunction, an engineer helped me record the data. Usually a young woman assisted in the recordkeeping, but I don't recall that we had such assistance on this particular night. All went well until about 11:30 P.M., when the air-raid sirens sounded. We were quite used to this unpleasant, penetrating noise, and thus had no fear that we would ourselves be the target of an air raid. Following the rules, we stopped the experiment and turned off the vacuum pumps, checked that no light was escaping the building, and stepped outside.

I recall that earlier in the evening it had been beautifully clear, with the full moon brightening the scenery. Now a blanket of artificial fog—a chemical cloud that obstructed the view of the ground from above—covered the area, and the moon had nearly disappeared from sight. Yet it was still rather light. We waited a while, and I don't remember whether I heard the sound of aircraft. Nothing seemed to be happening. We chatted, and I don't recall whether the all-clear signal was sounded, but at any rate we went back into the wind-tunnel building. Suddenly we heard the loud, cracking sound of an 88-mm antiaircraft gun, the type that I had seen in action against armored vehicles in Russia. Smaller guns, such as the rapid-fire 20-mm with which I had had experience, also began to fire. The characteristic hum of many large airplanes filled the air. We again left the building—this time in some haste.

Now the night was clear again. The artificial fog had drifted away from our area; the airplane noise became louder, and bombs began to explode some distance away from us at the center of the army station. It seemed quiet in the direction of the air force research and development station at the northwestern tip of the peninsula. The wind-tunnel buildings—and even farther out the thinly spread rocket test stands, firing areas, and related installations—were all located in the northwestern part of the army compound. Living quarters in the Siedlung, administrative buildings, mess halls, shops, the large pilot-plant manufacturing halls, and warehouses were more closely spaced some distance toward the southeast. It was in that direction that we heard explosions. It seemed to me, judging by the varied engine noises, that German night fighter planes had arrived. In addition, debris from the exploding antiaircraft shells began to rain down from the sky, producing a characteristic crackling noise when it hit leaves, roofs, or the ground. Clearly the moment to take shelter had arrived; a full-scale bombing raid against the army installation was obviously taking place.

The sandy soil of the Baltic coast does not lend itself to the construction of underground shelters. The groundwater level was barely below the surface, and therefore concrete cubicles with metal doors were located above ground in various places. (The test stands—not the wind tunnels—were built with heavy safety walls to guard the engineers and technicians in case of equipment malfunctions. Living quarters such as Haus 30 had full basements. The protection of people in such buildings was therefore about like that found in the cities.) A few of us walked to the nearest cubicle and closed the door behind us. The space, which was equivalent to a living room, easily housed our little group of three or four people. Others on the night shift went elsewhere.

I had of course experienced many air raids. In the Russian campaign there were attacks by the low-flying Stormovik airplanes, and we were also bombed frequently by small numbers of other airplanes. If one found a ditch or hole, the chances of being hit were relatively small, and the attacks did not last long, in contrast with the sustained, accurate artillery fire for which the Russians were well known. Primarily, however, one was not cooped up; one could see the sky and gauge what was going on. When we were in a village it was another matter: single older planes dropped an occasional bomb at night, often thrown over the side by hand. Being inside a house—usually built like a small log cabin—was unsettling. But I was really frightened sitting in the cellar of our house in Berlin during my second study leave early in 1943, when I experienced major air raids. My assignment on the roofs of Berlin during the winter of 1940–1941, after the French campaign and prior to the invasion of the Soviet Union, was comparatively harmless. A few British bombers appeared at night, and the searchlights and antiaircraft fire produced an impression of fireworks rather than real danger. At any rate, here in Peenemünde I was again locked up, but things in our little concrete shelter did not seem too bad at all. There was much noise, and at one point an unexploded flak shell or a small bomb hit a corner of the bunker, tilting the whole structure slightly. Still, it was apparent that heavy bombs were not falling in our area.

Sometime after midnight—I am uncertain about the duration of the raid —the noises abated. We stuck our heads out the door and saw fires burning in the southeast. A few airplanes were still flying about, apparently descending quite close to the ground. At that point we left our shelter. We ran toward the wind-tunnel building and soon noted in the moonlight that the structure stood as we had left it; luckily, no destruction had taken place. Noises of running and shouting people and the crackle of fires dominated the scene after the airplanes left. I don't recall whether the sirens ever gave the all-clear signal to inform us that the raid was over.

Between the wind tunnels and the nearest buildings in the direction of the living quarters were storage sheds and a stand of scrawny fir trees. Next to the sheds, a number of huge rolls of cable had been stored. Some of these rolls burned brightly, since their tarlike insulation was highly flammable. Our small wind-tunnel group began to push the rolls to a clearing, but by then the trees were on fire, and so we got tools and began cutting a firebreak to save our building. It is remarkable how much hard physical labor young people can perform under such circumstances. After the fire danger seemed to be over, I started toward the Siedlung, which included my wooden barracks and the old Haus 30. Fires burned everywhere, and confused people ran to and fro, but I do not recall seeing seriously wounded or dead people at that time.

Slowly the light of dawn appeared. Many of the houses had burned down by now; the wooden ones, like my quarters, were all gone, while stone walls remained. I tried to help here and there at various tasks. Major and minor fires burned out of control. Enormous confusion prevailed. Trucks carted off the wounded. Nobody seemed to have any idea of the extent of the damage, the number of people killed, or what to do next. The rail tracks were largely destroyed. The fences were ripped, and no guards were visible. Occasionally I saw somebody whom I knew, and at such moments both of us were outrageously happy to be alive. Nobody in all the highly structured hierarchies of the guards, the different army and air force units, or the civilian employees appeared to be asserting command and leading people to concerted action. Later I found out that help for the wounded and dying had been organized more efficiently than I knew. I had, in fact, not been at the real center of destruction.

During the early afternoon—I had been up for about thirty hours by that time—I felt there was little else that I could do myself. The wind tunnels had survived essentially untouched; my own things were gone in the fire, but I was alive and unharmed. Therefore I decided to make my way to Koserow. The single road parallel to the coast was clogged with trucks, staff cars, motorcycles, bicycles, and pedestrians in both directions, and all moved at a crawl. Hitching rides, I slowly made my way to the hotel. I cleaned up and walked to the beach, where I found my wife.

I now heard that she, in common with most people in Koserow, was rudely awakened in the night by the rumbling of falling bombs and anti-aircraft fire. She got up and viewed the raid from the distance. A bend in the coastline projects Peenemünde outward, so that the frightening spectacle could be seen in its entire extent. It appeared from Koserow that nothing on the ground could possibly survive this inferno. Some observers claimed that they saw aircraft being shot down, in addition to the explosions and fires at ground level. My wife told me that at dawn many trucks and cars carrying people wounded in the air raid passed by the hotel window. She decided to go to the wind tunnels to find me. Among the few cars moving toward the laboratory she found one with a general who took her along. She had no problem entering the normally heavily guarded installation, and she found the wind tunnels untouched. Although I was not around, she encountered someone who knew me and who had seen me alive and unharmed. She then returned to Koserow by hitching rides. But now I could barely keep awake, so I went to the hotel, ate, and fell asleep.

Later the next morning I hitched my way back to the laboratory. The brilliant weather held, and the sun illuminated the destruction clearly. Fires no longer burned, and people were moving in all directions to bring order out of chaos. We cleaned up in and around the wind tunnels, which indeed were as good as new. Soon power was restored in some areas; the power plant supplying Peenemünde was untouched, since it was far from the station. A strange mood pervaded the place. On the one hand, it was obvious what had happened in the area; on the other hand, a mixture of rumors, ridiculous stories, and factual accounts made the rounds. Why were the major test stands, the wind tunnels, the liquid oxygen facility, and the power plant unharmed, while the living quarters were destroyed? Were the lead planes confused by the location of the drifting fog? I have not yet found sufficient rational answers for such questions. Among the facts we heard, however, was that the air force station had not been hit, and in retrospect this omission by the British bombers seems especially strange.

After inspecting the wind tunnels and exchanging stories about our experiences with my friends at the institute, I went to where my room had been and poked through the cold debris. The remnants of a camera and an alarm clock emerged from the ashes; they were beyond restoration. Although this sort of thing causes an eerie feeling, the sense of loss of material belongings is negligible in comparison with the joy of being a survivor. While walking about and inspecting various sites, I saw the great Oberth stirring around in the ruins of a building. Occasionally he bent down and picked up something. I greeted him, and we talked about the air raid. He was collecting blackened nails, feeling that they would again come in handy.

One of the more frightening but unverified stories involved the residents of one of the large houses for young women who worked as secretaries and technical assistants. The group initially huddled in the basement that ran the length of the burning building. Mass hysteria set in, and the women broke out and fled to the beach. It was said that after the fighter planes disappeared and the flak was silenced, several British bombers flying at low altitude strafed the beach with machine-gun fire aimed by the rearward-facing tail gunners. Bodies were found on the beach. Because I did not go to the shore, I don't know whether this story is true; I do know, however, as noted before, that at the end of the raid some airplanes flew very low. Another story was told about two housemothers who were taking care of the younger women in another basement. Here all prayed on their knees during the raid, and the two iron-willed matrons guarded the doors and kept the panicky women from leaving the cellar. All were saved, even though the building above them burned down.

Wild stories circulated about the capture of prisoners, the appearance of spies, and an approaching daylight raid. It is noteworthy, however, that no daylight bombing occurred for a long time. It seemed to me then that a daylight attack by just a few airplanes immediately following the major air raid could have destroyed the technical facilities. It was a clear day and the antiaircraft guns were out of commission, which would have abetted a follow-up attack. But this thought again touches on British bombing strategy, a different subject altogether.

Many of the dead were collected in coffins, and a mass grave about the length of a soccer field was dug. At the funeral service for these victims of the air raid, the Flakversuchsstelle—the antiaircraft unit of people such as me assigned to work in the laboratories, test stands, and pilot-plant facilities—assembled for the only time during my tenure at Peenemünde. We were in uniform, lined up by the huge grave in a rather haphazard formation. Speeches, music, and religious rites—but little political pep talk—made for a moving experience. There was a sense of imminent change, too. The work of years at the station, in which I had taken part for only about three months, had reached a strange culmination.

I was fortunate to be in the relatively safe area near the Aerodynamics Institute during the night of August 17–18. But how did the center of the installation fare, and what do we know in retrospect about the raid and events leading up to it? Dornberger describes what happened to him, von Braun, and others who were in the middle of the action. Like me, Dornberger was at first not perturbed by the screaming of the sirens at about 11:30 P.M. The blackout was checked and found to be in order; the full moon shone brightly. Dornberger called the defense center and was told that the course of the approaching airplanes was not yet known, so he went back to sleep. Soon he was awakened by the flak firing from their positions at the airfield and the settlement. Windows broke, tiles fell off the roof, and he got up hurriedly.

Outside he found the artificial fog blanketing the area in a dense, rose-colored hue reflecting the first fires started by the bombardment. The characteristic low-pitched noise of masses of four-engine bombers filled the air. He raced to find von Braun, who said that the raid was the real thing. Together they rushed to a shelter, where they were joined by many scared people in night clothing. Telephone connections were ruptured, but Dornberger succeeded in getting in touch with the air-raid defense commander. He found out that shortly after midnight the first airplanes had flown to the south in the direction of Berlin, repeating the normal pattern of past nights. Soon, however, the airplanes turned, and wave after wave of heavy bombers appeared over Peenemünde, dropping their loads. So far no damage to the test stands and other facilities spread out farther north had been reported. But major fires broke out all over the center of the army installation; the assembly hall burned, additional fire engines from nearby locations had to be called, and the last doubt that a major attack was in progress disappeared.

Dornberger and von Braun left the shelter after the worst appeared to be over. Sparks from the fire flew through the air; cans of phosphorus burned on the ground. In this inferno they tried to organize rescue missions to preserve the safes containing drawings and calculations. About one and a half hours after the first bombs were dropped, the flak fire ceased. Low-flying aircraft appeared over the beach and indeed strafed the area. It was more than two hours before messengers reported details on the fate of the settlement, where several thousand German civilian employees lived. About half of them were scientists, engineers, and technicians. Nearly all the buildings were destroyed, and reports of mounting numbers of casualties were coming in. It became known that among the many dead were Walter Thiel and his family, buried in a ditch. This was the first news of the death of one of the most important technical men. Thiel was a pioneer in rocket development long before his appointment at Peenemünde, and his major contribution was the development of the rocket engine itself.

Apparently the air force station was unscathed. At the army laboratories, no serious damage was reported in the main assembly plant, where the experimental A4 was to be turned into a long-range missile that would be mass-produced. The rocket test stands and firing facilities—just like our wind tunnels—were practically untouched. All early messages dwelled on the disaster in the settlement and in the central area of the station. At this point in his description of the raid, Dornberger records his thoughts on the senseless destruction, the killing of old folks and children, and the terrible effects of the air war as manifested in this raid. It is odd that he does not connect the sad events at Peenemünde with the plans for the bombardment of London and other cities in England with the A4. Later in the day, Dornberger and von Braun flew in a small plane over the army research center, and Dornberger exclaimed, "My poor, poor Peenemünde!"

It took weeks to determine the total number of dead, including those who died from their injuries after the raid. Dornberger cites 735 victims; the same tally is given by Oberth's biographer Barth and by Klee and Merk. Stuhlinger and Ordway report that about 800 lives were lost. By far the largest number—perhaps as many as 600—died in the closed camp called Trassenheide. Here a large number of prisoners of war and foreign construction workers lived under prisonlike conditions. Among them were Polish and Soviet citizens, some of whom may have volunteered for Peenemünde to escape even more unpleasant prisoner-of-war camps. The Trassenheide inmates were fully exposed to the fury of the bombing in their heavily fenced area.

It is now known that the underground grapevine of Polish laborers in the camp had long been producing precise bits of information on Peenemünde, as noted by Johnson.[1] Early in 1943 "heavy, vibrating noises" were reported, and sometimes prisoners had glimpsed a small airplane "with a flaming tail" flying out to sea. Obviously the Kirschkern, which I noted at about the same time, had been discovered. These stories reached the Polish general staff in London in the spring of 1943. Allied intelligence agents in Germany had also passed on messages that something unusual was happening on Usedom. Johnson notes these events and also cites an earlier discovery of mysterious goings-on: on May 15, 1942, a Royal Air Force flight lieutenant was sent to take reconnaissance photographs of the Baltic seaport of Kiel. After completing his mission he had a few shots left, so he made a pass over the nearby airfield of Peenemünde. The pictures did not show anything highly unusual, but they revealed "elliptical earthenworks." Since the British evaluation team had not been asked specifically to look for these strange structures, the first photographs of the A4 firing stands were duly numbered and filed away!

It is further reported in the sources just cited that as early as October 1939 an anonymous letter written in German reached the British embassy in Oslo. The Peenemünde activities were described, and accurate details of the planned A4 were given. The British secret service supposedly ignored this information because the letter was regarded as a diversion cooked up by its German counterpart. Evaluating such stories is better left to historians; it is certain, however, that in May 1943 Flight Officer Constance Babington-Smith of the Allied Photographic Unit in London noted on an aerial photograph a small airplane without a cockpit for a pilot. This photograph, together with later photographs of a "strange cigar-shaped object" and many reports of new and potentially threatening secret weapons being developed on the Baltic shore, finally resulted in action by the War Office Intelligence Branch. The thirty-five-year-old Duncan Sandys, a son-in-law and confidant of Winston Churchill, was assigned to investigate the possibility that long-range rocket weapons were in fact under development. He was soon convinced by the mass of material that such was the real function of the Peenemünde laboratories. Consequently, plans for a massive air raid were prepared.[2]

In parallel, German intelligence formally warned Dornberger of possible air attacks. Dornberger himself was already convinced that the story was out. One of his employees had discovered a secret message in the solution of a crossword puzzle printed—innocently, one must assume—in a German illustrated magazine. More directly, Dornberger noted that what he called "frozen lightning," the zigzag signature of the rocket exhaust, which was visible in the sky for large distances in good weather, could not be concealed.

This brief summary of the events leading to the discovery of the purposes of the air force and army stations at Peenemünde is sufficient to allow me to contrast the reality with my own thoughts then and now. For one thing, I find it difficult in retrospect to understand that I never seriously contemplated a possible air raid against such an attractive target. How could I have been so simple-minded in view of my experiences in the war, especially during the air raids on Berlin and the frequent alerts at Peenemünde? Apparently my delight in the altered life-style kept me from pondering the future of the laboratory. The second retrospective puzzlement relates to the Allies' delay in recognizing the nature of the work at Peenemünde. After all, the establishment was founded prior to the war, and major construction activities took place while the Usedom beaches teemed with vacationers. Travel was open to all: Americans could easily have visited the area as late as December 1941. I imagine disguised spies with binoculars strolling along the seacoast. Moreover, all the efforts at the laboratories took place while hordes of technical and political visitors (see chapter 4) arrived with their entourages, and they were shown whatever advances had been made.

At any rate, the night of August 17–18, 1943, became the date of reckoning. In all, 597 British Halifax and Lancaster four-engine night bombers—accompanied by 45 lead aircraft and night fighters—dropped about two thousand tons of explosives over Peenemünde. About forty aircraft and 240 men did not make it back to their bases. It will take further research to learn why the settlement was destroyed while the technical facilities remained largely intact. After all, the raid took place under a bright moon, and the artificial fog had largely been blown away. The bombing strategy employed against large cities appears singularly inappropriate for the Peenemünde raid. Later daylight air raids—on July 18, August 4, and August 25, 1944—long after I had left Peenemünde, were more effective. Their scope, however, was much smaller than that of the night raid that I experienced.

We shall see in the next chapter that the further development of the A4 was little affected by this event, which led only to a modification of plans on the German side. The later implementation of these plans had important long-term consequences for my own life.

SIX

The Aftermath and the Move to Bavaria

T HE SMOKE HAD CLEARED, THE DEAD WERE buried, the debris was accumulated in piles, and things such as a toothbrush and clothing had been procured to replace what had been destroyed. The actual damage could now be surveyed. The effect of the air raid, while smaller in overall scale, was similar to that seen in industrial cities in Germany. Laboratory and testing facilities were relatively unharmed, just like the factories of the cities. But administrative buildings, mess halls, and housing were in large part destroyed.

Dornberger notes these facts, and he estimates that no more than six weeks were lost before work could be resumed in full force.[1] However, the appearance of destruction was carefully preserved to mislead aerial reconnaissance. The picture of devastation was even enhanced through camouflage. No further air raids occurred, and the fiction of destruction was kept up for about nine months. These circumstances were not known to me at the time.

Clearly, the discovery of the birthplace of large rockets and the subsequent air raid came too late to scuttle or even seriously hinder the progress of the various projects. In fact, the first start against London of a flying bomb, the air force's V1, was barely delayed. In any case, the army's A4—soon to be called the V2—would be ready after the V1, owing to its vastly more complicated design and difficult handling in the field. In addition to its liquid fuel (75 percent ethyl alcohol and 25 percent water), the A4 had to be filled with liquid oxygen to provide a combustible mixture for the rocket nozzle that propelled the missile. Liquid oxygen is a dangerous, extremely flammable substance that must be kept at a very low temperature. In contrast, the V1 was in all respects other than its intermittent engine a pilotless airplane that was started from an inclined ramp, the type seen in the British reconnaissance photographs.

The conversion of the experimental A4 to a missile that could be mass-produced like other military hardware was far advanced. Consequently, it was decided to push the project quickly to completion, and to do it at Peenemünde, using the largely untouched facilities. The electrical power for such work existed after the transmission lines had been repaired. This speedup implied that bugs in design and manufacture would have to be rectified afterward.

The final large-scale production factory was to be housed in a defunct mine in the Harz Mountains, in the middle of Germany. Construction work on this plant had started; it was now to be sped up as well. We were approaching the winter of 1943–1944, a time when even simple items began to be scarce. Given these conditions, it is noteworthy that such an unusual undertaking as the construction of an underground mass-production factory for radically new weapons could be even contemplated, let alone carried out rapidly. It is important to remember that rail transportation still functioned reliably, albeit slowly. The continually wavering Hitler decided to lend his full support to these wonder weapons. Had he reached a state of desperation? Whatever the answer, his support set in motion an ominous development. Heinrich Himmler—the head of the SS and the Gestapo (the secret police), who had previously been briefed at Peenemünde—set the stage for a takeover of all rocket development, a move that was sensed by Dornberger and von Braun, who became most apprehensive.

Meanwhile, regular testing could be resumed immediately in the wind tunnels, and the Wasserfall experiments took place in late fall. Even as work was progressing, however, it was evident that there was no reason for the wind tunnels to remain in Peenemünde. After complicated deliberations that I was not aware of, the decision to move to Kochel was made. A lovely, picturesque town, Kochel is located literally at the foot of the Alps, about sixty kilometers south of Munich, at the end of a railroad spur. To me the reasons for the move seemed obvious; however, I discovered as late as October 1991 that a serious tug-of-war between von Braun and Hermann had preceded the decision. That internal disagreement caused major difficulties.

At this juncture marked by the departure of the Aerodynamics Institute from Peenemünde, it is interesting to review some of the wind-tunnel work that had been carried out.[2] The external aerodynamics and the configuration of the graphite control surfaces in the rocket jet of the A4 worked well, as demonstrated in several successful test flights at Peenemünde and in Poland. The A4 was close to mass production and deployment. In contrast, the development of the antiaircraft missile Wasserfall moved at an erratic pace.

Only in 1944—as I know now—under a crash program to break the so-called *Luftterror* (air terror: by now air raids were classified as attacks by terrorists), did the Wasserfall, together with other truly eccentric schemes to shoot down airplanes, obtain a high priority. The aerodynamics of the Wasserfall were well in hand, except for the details of the final shape of the control surfaces attached to the rear fins. I was still involved with this matter. I could never understand, however, how a rocket flying at three times the speed of sound could follow an airplane that flew at low speed on an unpredictable course. This seemed difficult to me not only because of the different speeds of missile and airplane, but also because of the different schemes of homing in on an aircraft. I had heard of two such schemes, one being the optical method of sensing the heat of an airplane's internal-combustion engines, the method that Jordan worked on. The other method was to direct a radar beam on the aircraft and provide the Wasserfall with electronic devices to fly continually glued to this beam.

Aside from the rocket work of high priority, many less-extensive test series unrelated to the goals of the Peenemünde labs had been carried out in the wind tunnels. Such work was assigned directly by army headquarters in Berlin. Similar extra experiments had to be expected, wherever the Aerodynamics Institute might be located. Research on models of artillery shells, which were spun by small electric motors in the test section to simulate real shells, and on models of strange, high-speed bombs, small rockets, and the like was carried out by Lehnert's group. I was preoccupied with the design of the control surfaces and knew little about the many other projects that used the wind tunnels, including those completed prior to my time. However, one idea that appealed to the classical artillerists deserves discussion. A scheme to vastly increase the range of existing cannons had been proposed in-house by Hans Gessner, an imaginative engineer. Gessner supervised the mechanical design and construction of basic facilities, including the planning on the hypersonic wind tunnel to be built at Kochel. At the same time, he invented an artillery shell that was stabilized aerodynamically rather than by the gyroscopic effect of spin impressed by rifling in the gun barrel. His finned shells had to be quite long to fly on an aerodynamically stable course, and they needed to be guided in the barrel by so-called sabots. These disk-like inserts centered the shell in the barrel, and they dropped off after leaving the muzzle. Such new shells were indeed built and tested on firing ranges. The radically new idea gave the projectile a much longer range, with total weight and explosive charge equal to that of conventional ammunition. In recent years, reports of supercannons have appeared in the press, and Gessner's invention may explain one aerodynamic aspect of such monsters.

Kochel had long been selected as the site of the 1-square-meter hypersonic wind tunnel, which was to operate to a Mach number of 10. Land was now purchased (or requisitioned?) at the edge of the village, at the foot of a steep slope. While the equipment was being moved from Peenemünde, buildings for the machinery, the wind-tunnel test sections, and the laboratories, as well as wooden buildings for offices, were being erected starting in December 1943. A small factory nearby was leased for use as a machine shop and as housing for technicians and model designers. All this frantic activity had to take place in secrecy. The Aerodynamics Institute at Peenemünde had, of course, been attached to the administration and logistics of the immense complex of the army. At Kochel the funding came directly from Berlin, and the fiscal management, personnel offices, purchasing, security, and fire protection had to be added from scratch. Several administrators were hired for this purpose, and the total staff swelled precariously. At the height of the activities in Bavaria, substantially more people than at Peenemünde—many with families—were involved in the operation, and all had to be put up in the already crowded small town. This created serious problems with the local population.

The institute and its site were formally incorporated under the name Wasserbau Versuchsanstalt (WVA). This camouflage designation suggested a civil engineering research institute concerned with the design of dams, weirs, locks, and canals, all of course handling water. Ironically, not far from Kochel in the mountains was a real WVA, leading to additional mix-ups. Hermann was designated as director; Kurzweg and a newly hired administrator, Herbert Graf, were the assistant directors. Eber was the head of planning for the hypersonic facilities.[3]

Official mail from Peenemünde, Berlin, and other places was received at a post office box in Munich. Every day a mail courier was dispatched by train or motorcycle. The business with all local contractors and suppliers was handled in Kochel. I found out after the armistice from a British intelligence officer that Allied intelligence did not discover the location of the Aerodynamics Institute—that is, the WVA. It was understood that a post office box in Munich was our mail drop, but the location of the tunnels remained a secret.

Those in the military like me (I was still a member of the Flakversuchsstelle) remained assigned to the wind tunnels. Fortunately, we had to wear civilian clothes, and we were forbidden to divulge our affiliation. No military presence was ever noticed in or around the institute. Our total freedom outside working hours and the scenic surroundings made for a pleasant life in a rapidly deteriorating country.

Slowly the equipment was transported from Peenemünde to Kochel. During a period of about one year, roughly three hundred freight cars loaded with the wind tunnels, the vacuum pumps, the drier, electrical equipment, entire laboratories (optics, photography, model building, machine shop, and so forth), furniture, and private possessions arrived at the small railroad station at Kochel. This station today looks exactly as I knew it during the war, though the town is now a thriving vacation spot overwhelmed by automobiles.

The transfer of the wind tunnels was made in two stages. First, one of the two large test sections was moved while the other remained in operation at Peenemünde, primarily for experiments with the high-Mach-number nozzle. To allow simultaneous testing at the two sites, a cylindrical vacuum tank of 750 cubic meters, rather than the original 1,000-cubic-meter sphere, was built at Kochel. This smaller vessel permitted only slightly shorter blasts for the aerodynamics experiments. Moreover, to retain the Peenemünde operation, one of the three sets of pumps was left there. Although the complete transfer of the Aerodynamics Institute—now independent of Peenemünde—took about a year, the first tunnel began to operate in a little over half this time. The test section left at Peenemünde was dismantled in May 1944 and moved, and full-fledged testing with both wind tunnels resumed in November of that year. The intermittent hissing noises of the tests issued from the thin-walled structure surrounding the test sections and reverberated between the sides of the mountains.

Rumors of dangerous underground installations sprang up among the locals and the displaced; wild ideas circulated, and people worried about prospective air raids on a secret war plant. It had seemed bad enough before, when a flood of northern refugees arrived in Kochel to live next door to the largest German hydroelectric power plant. Now a secret weapons factory had been added, whose strange noises compounded the confusion. It is still puzzling to me that apparently nobody—including the local population—ever discovered exactly what went on. After all, trains arrived with heavy machinery, substantial local construction took place, and many families appeared and sent their kids to local schools. Something ought to have alerted people to the rather simple truth.[4]

All this, including in particular the appearance of additional folks from the strange and historically unloved German north—Prussia—caused consternation and worry among the locals. In 1943 the little town was already jammed with refugees from the bombing attacks on northern cities. I believe that about six thousand people resided there at the end of the war, whereas the usual population was roughly fifteen hundred. The first newcomers were women and small children evacuated to escape the siege from the air. Entire schools, including their teachers, next moved into the hotels of Kochel. Owing to its beautiful location, the town's primary source of income was normally tourism. Since tourism ceased during the war, much space was available in Kochel initially. But when the crew of scientists, engineers, and technicians arrived in 1943, living space was tight indeed. And many of us, including the number that were called "Herr Doktor," had trouble communicating with the local populace because of the dialect differences. The distribution of food through ration cards was still relatively generous, but the small country stores were overtaxed, and many purveyors of food took advantage of the new crowd. All this opened a fresh chapter in the varied history of a town located on one of the standard routes to Italy—one traveled, for example, by the young Goethe in 1786.

It was decided that I should stay on in Peenemünde to the end of the wind-tunnel operations there. I assume that the Peenemünde authorities hoped to see the end of work on the aerodynamics of the Wasserfall. What I have reported here about the initial Kochel activities is therefore based in part on the reports of travelers and the cited sources.[5] I continued to work on the Wasserfall, carrying out, for example, the elaborate series of experiments on pressure measurements described in chapter 3. Little official work was required after these experiments, and we had the glorious opportunity to try this and that, just as we had in a later period toward the end of the war. Virtually any experiment in basic fluid mechanics at supersonic speeds was new. For example, we measured the aerodynamic drag of a sphere in a range of Mach numbers to compare the results with well-known low-speed data and supersonic experiments with spheres fired from guns. Not only were such results new to us, they did not exist anywhere else, though I didn't know that at the time.[6]

Erdmann, in the meantime, prepared a final grand experiment before the last tunnel was to be dismantled. More or less on his own initiative, he wanted to achieve a flow of close to nine times the speed of sound. His design to achieve this goal was actually superior to the design plans of the new hypersonic tunnel at Kochel. I witnessed only some of the preparations for Erdmann's experiments. He intended to replace the entire test section of the remaining large tunnel with a special box that could take high pressure and house a nozzle designed for $M=8.8$. The design was based on experience that had served well up to $M=4.4$. The resulting shape looked unusual. An extremely narrow slit at the throat—the narrowest part of the nozzle, where the speed of sound is attained—was followed by two symmetrical, sharply opening nozzle walls, again enclosed in parallel plate-glass sidewalls. Erdmann—unlike those who planned the large hypersonic tunnel at Kochel—correctly anticipated that a hypersonic nozzle must be fed with dry air at a substantially higher pressure than that of the atmosphere.

Supersonic Mach numbers, as I noted in chapter 3, are produced by an expansion of the air beyond the sonic throat in the diverging part of the nozzle. This expansion is a process in which pressure and temperature fall simultaneously. At high Mach numbers, expansion leads to lower and lower pressures *and* temperatures in relation to the initial conditions ahead of the nozzle. For the test Mach number planned by Erdmann, the ratio of pressure ahead of the nozzle to that in the test section was in fact over 18,000. Therefore, a high pressure was needed at the starting point of the expansion ahead of the nozzle to avoid a condition close to a vacuum in the test section.

It was well understood that an expansion of dry air starting at room temperature would lead to very low temperatures in the test section at Mach numbers above 5 or so. A group at the Technical University of Darmstadt led by the physical chemist Carl Wagner was under contract to Peenemünde to study this problem. Nozzle flows were computed to find out whether the highly cooled air—that is, the nitrogen and oxygen making up the air— would condense, much like the water vapor whose removal had necessitated the introduction of an air drier. There seemed to be a chance, however, that the air might *supersaturate* and not change into a fog of liquid droplets or solid particles. Supersaturation would be achieved if the air were to expand beyond the point where it is supposed by the textbooks to condense, without changing to a liquid or solid "air" fog. A reasonable expectation arose that if air were cooled very rapidly it might not condense, and aerodynamic testing would be possible. At any rate, this sort of thing was totally unexplored, and experiments were required to verify or disprove the underlying theory.

Everything was set for the first-ever hypersonic flow experiment. The highest possible pressure ratio across the test section was achieved by evacuating the sphere to the limit the remaining pump could achieve. The supply of the nozzle—in contrast to that at lower Mach numbers—was now provided by air at a pressure of about 90 atmospheres, that is, a pressure ninety times higher than that of the atmosphere surrounding us. The experiment was initiated by opening the fast-acting valve. The flow of brief duration looked perfect as viewed via the optical system. Beautiful photographs of the flow about wedge-shaped models, cylinders, spheres, and other simple shapes were taken, photographs that looked just as one would expect from gasdynamics theory. The measurements of pressure were difficult under the circumstances; however, they too seemed to indicate that the design Mach number of the nozzle had been reached.

One flaw became apparent. Looking at the flow through the glass walls, one could see a dense fog. We now know that under the conditions of this particular experiment, the air had indeed partly condensed. The fog was made up of air droplets or solid air particles forming a cloud, just like the water clouds we see in the sky.

Little did I anticipate that my marginal role as a distant onlooker at these pioneering experiments would lead to a lifelong interest in such condensation phenomena. It wasn't until about 1950 that an experiment like Erdmann's was fully understood, and I had a part in this later development in the United States.[7]

During the winter our experiments proceeded slowly since we had only one remaining pump, a situation that led to long intervals between runs of the wind tunnel. In fact, I do not remember any striking events that occurred during this leisurely period. I did have many occasions to learn more about all aspects of the wind-tunnel operation, ranging from science to an understanding of what can and what cannot be built in a machine shop.

I received short leaves to see my family in Berlin, and I still traveled occasionally as a courier. In the capital, life had returned to noteworthy normalcy after the disastrous series of air raids earlier in the year and the evacuation of children and others in August. (Additional major air raids on Berlin occurred after I had left early in 1944.) This normalcy existed in the face of nearly nightly siren blasts signaling smaller attacks. Berliners are well known for a somewhat cynical sense of humor that served them well during the war. It had become clear to almost everyone that the war was going badly. The daily activities of life were becoming increasingly difficult: food rations were slowly decreasing, clothing was in extremely short supply, soap was hard to find, and much time was taken up with waiting in line. Slowly the Germans had to adjust to the supply situation of the rest of German-occupied Europe. Theaters, movie houses, and other places of entertainment were crowded. They operated almost normally since they were strongly supported by the government, a situation that was similar to the availability of lipstick! In short, the adaptation to a severely altered life-style was reasonably successful. In retrospect, as I have already said, I am puzzled that I never expected another air raid on Peenemünde. In fact, none occurred until after I left for Bavaria in the spring of 1944, when American daylight raids began.

At some point during this period Erdmann asked me what I thought of the pace at which the institute was being moved to Bavaria. At the time, I felt that things were moving rather slowly, and I said so. In retrospect, I can see that this statement demonstrated only my lack of experience. Here we were in the winter of 1943–1944, in a major laboratory that had been damaged by bombs, and we were to remove the undamaged Aerodynamics Institute as rapidly as possible so as not to lose precious testing time. While some of us were still at the Baltic, the new institute near the Alps was in fact quickly taking shape. I would later consider the move a miraculous feat. An entire high-technology institute with delicate machinery, machine shops, electrical shops, optical laboratories, books, documents, furniture, and office supplies, as well as a diverse group of people with families, was moved from the northern border of Germany to its southern border in the face of continual air raids, slow transport, and a rapidly decaying economy. New buildings had to be erected in spite of extreme shortages of building materials, new people had to be hired to take responsibility for the services that had been provided by the central administration of the army facility, and all this needed to be carried out in complete secrecy. After the subsequent move of the wind tunnels to the United States in 1945–1946—which had full support, practically unlimited funds, and the labor of well-fed people who were not bombarded from the air—it was much longer before the first wind-tunnel blast shattered the air outside Washington, D.C.

The background of Erdmann's question about the speed of the move was not explained until 1991, when I visited him in Holland. At the time of our discussion in 1943, he had prepared a *Denkschrift* (a strategic memorandum) that he planned to submit to Wernher von Braun, who was his friend, and the leading authorities of Peenemünde, bypassing Rudolf Hermann, his boss and the director of the Aerodynamics Institute. He proposed leaving one of the large wind tunnels in Peenemünde rather than moving it to Kochel after the completion of the high-Mach-number tests, as currently planned. Since it would take some time to get going at Kochel, under his plan aerodynamics testing would be continually available at Peenemünde in case aerodynamics problems arose with the A4 or the Wasserfall. The proposal contained a complete organizational separation of the two aerodynamic operations, a separation that would cut off Hermann from his Peenemünde connection. Possibly because Erdmann had suggested that Eckert and I stay with him, I underwent an interrogation on my views concerning this point. I was questioned by people at Peenemünde whom I did not know and could not place; I was worried at the time, though I cannot remember more details. Fortunately, this was the last I heard about this matter at the time.

Rumor mills operate efficiently in a small closed society like that of the rump Aerodynamics Institute, whose numbers had by now shrunk to the point where hardly any formal structure remained. It was said that the SS was involved (not true), and other wild ideas were associated with the Denkschrift. In retrospect I understand that Hermann was strongly opposed to the truncation of his institute. He fought the proposed split and won; Kochel became the only aerodynamics center for the army, and Hermann's position was not diminished. His success affected my future, and I followed my colleagues to Bavaria. I much preferred this move in view of Peenemünde's exposure to the Soviet advance.

The year 1944 arrived. A letter written during the night shift informed my mother on March 4 that I would spend a last weekend in Greifswald. Since my wife studied there, we had met a number of professors whose company I much enjoyed. I said good-bye to a professor of Romance languages whose major occupation at the time was a study of what really happened between George Sand and Alfred de Musset during their joint sojourn in Venice. Nothing could have been further removed from the actual conditions of Germany and the state of the war. I was delighted to have found one person who worked on something detached from current events. After this weekend I would get ready to leave Peenemünde for good, and my wife and I would travel by train through Berlin and Magdeburg to Essen, where my wife's parents lived. After that, the journey would take us to our old boarding school on the banks of the Weser River. Next a stop at Heidelberg was planned, and the final goal was Munich. I could not tell anyone that the actual final destination was Kochel, of course.

Rail traffic moved slowly; however, the trains did get you where you wanted to go sooner or later. Night travel took place under blackout conditions, and distant flashes of detonating bombs could be seen. It actually took twenty-four hours (not counting a visit in Berlin) to reach the city of Essen in the heart of the Ruhr area, the home of the Krupp works. The densely packed large cities of the Ruhr, the heart of German heavy industry, had been extensively bombed. Rubble was visible everywhere. However, the factories, mines, steel mills, and the like located between the population centers appeared to be largely untouched. The train passed plant after plant with belching smokestacks. I repeat that we know now that German industrial output peaked in the fall of 1944. The distribution system used to provide for the civilian population seemed to crumble earlier. Other than Peenemünde, Greifswald, and Berlin, I had not seen much of the country in some time. Now, however, observing people on station platforms, fellow passengers, and railroad employees, and listening to rather open conversations, it became obvious that times had become tough indeed. Food, clothing, and household goods were at a minimum, and people looked depressed.

After a stop in Essen we visited our boarding school, an oasis at the edge of one of the great forests of Germany. One could have open conversations with most members of that special community. Surprisingly little knowledge of the state of affairs was shown by my former teachers and friends that I spoke with. Some still believed in the final victory. There was of course one momentous fact that had touched all: the large number of deaths of students from the school at the various fronts. And they were so young. The last school year before the *Abitur*—the final high school examination—had long since been lopped off; in fact, my class in 1936 may have been the last to experience a full curriculum. I heard of many of my school mates who had fallen.[8]

Next our cumbersome travel took us to the beautiful, untouched city of Heidelberg.[9] Already rumors had sprung up that it would remain safe from bombs because the American army wanted to use it as a headquarters town. Finally we reached Munich and changed to the local line running to Kochel. Here the trains were electrified: the power plant at Kochel had remained untouched, and plenty of power was available. I had to shed my uniform because, as I mentioned earlier, the wind-tunnel camouflage required civilian clothing for those of us in the service. I recall the strong feeling of anticipation of what must clearly have seemed to be the last phase of the war. We expected the Allied invasion in France, but I did not realize how long the war would last. At least now we were far from the Soviet army and did not expect any air raids.

Leaving the train at the Kochel station, the end of the line, I found the local scene at first glance identical to that experienced at the start of a vacation. Snow-capped mountains, meadows and woods, traditional Bavarian houses, pleasant-looking inns, and people strolling leisurely in the streets. I needed a few days to settle in after reporting to the wind tunnels, and then reality took over with the discovery of the workings of the town that became my home—with interruptions—until the fall of 1945, several months after the end of the war.

SEVEN

Aerodynamics in the Mountains

MY FIRST IMPRESSIONS ON ARRIVING AT THE small town of Kochel on a clear day—not necessarily typical of Bavarian weather—must be further described. Long before the train from Munich reached the small terminal marking the end of the rail line near the Kochelsee, the snowcapped mountains became visible. A steep ridge called the Kesselberg separates the Kochel lake from the upper lake. The upper one is the large Walchensee, a beautiful body of water resembling a Norwegian fjord, with the mountains rising abruptly at the water's edge. The proximity of the Kochelsee, about 200 meters below, led to the construction of Germany's largest hydroelectric power plant. This type of plant is particularly suited to responding rapidly to fluctuations in demand, because the water driving the turbines can be turned on and off like a faucet. Transmission lines spread to the north, and with the increasing destruction of fossil-fuel power plants by air attacks, the Kochel plant gained in importance. Ultimately, the consumption of electric power was much reduced owing to the destruction of factories and the electrified rail lines, and thus we had practically unlimited power in Kochel.

The lake next to the village is bounded to the south by steeply rising foothills. A narrow path between lake and hills invites scenic walks. An old monastery in Schlehdorf, a village not far from Kochel, adds further interest to the scene. The area was untouched by war (excepting some artillery fire in the last hours before the Americans arrived); the deceptive bucolic charm reminded me of earlier summer vacations in Bavaria, and it hid the problems of the natives, the displaced, and the team of the WVA, the front organization for the wind tunnels.

The WVA arranged through the city hall for my wife and me to be assigned to quarters in a large private villa. The house was above the lake, shielded from view in all directions. The relatively new building was designed in the picturesque local style that was somewhat artificially varied to fit modern needs. Owing to rigorous zoning laws that are still enforced in most of Bavaria, no buildings of a substantially different appearance from those already built were permitted. The obviously wealthy owners of the house had so far fended off the housing demands of the many newcomers driven to Kochel by the war. The reason for the retention of an oasis of splendor amid overcrowded living quarters quickly became apparent. The place was owned by a high-ranking Nazi official active in Prague in the economic administration of German-occupied Czechoslovakia. I never met him, but his wife—our landlady—was most unpleasant. Our attic room could be reached only by lowering a trapdoor and climbing a movable extension ladder. But the view from a window under the eaves facing the lake was stunning and by itself more than compensated for the lack of hospitality. We saw the Herzogstand—a part of the first mountain range of the Alps—and the long ridge leading to the slightly higher Heimgarten, whose summits were a little less than 1,800 meters (6,000 feet) above sea level. This scene was framed by tall fir trees, and the lake could be seen below. It was certainly the most dramatic vista that I had ever seen from my own window. The view became even more satisfying after we had climbed the Herzogstand, traversed the ridge to the Heimgarten, and got into some trouble during the steep descent on our return to Kochel.

A short walk from the house toward the center of Kochel brought me to the WVA, where I shared an office with my friend Hans-Ulrich Eckert in one of the new one-story wooden buildings. Hans-Ulrich was a northern German who was disguised as a civilian just like me. As mentioned before, we could speak freely with each other, sharing our thoughts about the war, the actual state of the country, and Hitler during the period that was obviously the final phase of the war. In those uncertain circumstances, it was comforting to be able to talk to a friend about such questions as whose soldiers would first arrive in Kochel: the Soviets from Prague, the French, or the Americans.

Some rudimentary wind-tunnel operations started soon. Both of us struggled with the final design of the control surfaces of the Wasserfall. We kept changing the shape of the little rudders and again succeeded in making life tough for the machine shop. In addition, we did experiments on our own to enlarge our knowledge of supersonic flows. The missile work came to an end with our decision that we had reached a possible solution, a decision that was never supported by a full-scale experiment. Strangely, the "final" shape, named Rudder 21, was one of the items developed by my group that found its way into the literature on Peenemünde.[1] In the meantime, the last vacuum pumps and drive motors arrived from the north. Toward the end of the summer of 1944 the old routine was reestablished, with a break in the wind-tunnel operations of only about half a year. There were, of course, some changes, primarily in the work load. I do not remember that we ever had a night shift, and the whole operation slowed. We had leisure to take many trips by bicycle or on foot to the mountains and surrounding villages. I particularly remember an old monastery in the village of Benediktbeuren, not far from Kochel on the way to Munich. Here I saw a tablet with the names of a number of monks who had fallen in the war.

The ways of the Bavarians—who resented the most recent, and least predictable, bunch of newcomers—took some getting used to. One day my landlady praised the Finnish people's gallant resistance to the Soviet army, claiming that she would soon emulate their actions. I thought she was talking about a Russian invasion of Bavaria, an event certainly not unthinkable prior to the D-Day landing of the Allied troops in France. But she was thinking of defending her house and property against the likes of me, the invasion of the unloved Prussians. Since she still had several unoccupied rooms in her large house, she had every reason to think that the town might force her to quarter additional people.

The fat woman in the one dairy store used a big knife to cut butter. She always missed by a little bit the amount specified by the ration cards, and then she cleaned the knife by scraping it on the rim of an earthenware jar. Soon plenty of butter had accumulated, which she could sell for astronomical sums on the black market. The farmers traded food for Leica cameras and Persian rugs, with the cost rising as the expected end of the war approached. Nobody dared to discuss publicly the course of the war or anything that contradicted the official news. While many in the population were early followers of Hitler and had joined the party prior to 1933, nobody ever seemed to consider their role in the disastrous outcome of their initial hopes and ambitions. Certainly they did not view the Prussian invasion as in any way related to the war.

I vividly remember a run-in with an old party official who was the mayor of a small village near Kochel. One beautiful day I was cycling "illegally" while sirens were sounding to warn of an imminent air raid. That danger was indeed real for Munich, while the countryside always remained untouched. The old man stopped me and ordered me to get off my bicycle. When I refused, he pulled me off my bicycle and formally arrested me. At the police station, I was forced to show my Soldbuch. The local policeman let me go without asking me to explain why an air force officer in civilian clothes had nothing better to do than enjoy the countryside. My second arrest in the war ended without serious consequences—just like the first arrest at the railroad station in northern Germany. All in all, it was puzzling to see the powerful resentment toward outsiders of those who were the primary supporters of Hitler in Germany before and after the failed uprising in Munich in 1921, long before Berliners really knew what was going on down there. As I recall, shortly before Hitler came to power, my father told me that he had first read about Hitler in the newspapers in conjunction with the Bavarian coup. He saw Hitler's photograph, decided that the man was crazy, and promptly forgot all about him until the Nazi party became powerful in the early 1930s even in Berlin, the most cosmopolitan city in Germany.

Little of any technical interest happened in that period, but I do remember that I was always conscious of my luck in living in such tranquil surroundings. While we counted literally hundreds of American bombers flying north from their Italian airfields after crossing the Alps, not a single plane ever attacked Kochel, or more specifically the power plant. A rumor—not unlike that concerning Heidelberg—sprang up that the American forces wanted to preserve the hydroelectric plant to have a secure power source during the occupation of Bavaria. In fact, by now everybody assumed that the American army would be the first to arrive in Kochel, but it still was many months before this happened. Although hardly anyone noticed, once the war did come close to Kochel when at night a bomb fell into the Kochelsee. This exception was surely accidental: possibly an aircraft dumped its last bomb on the way back to base.

Aside from mountain hikes and cycling for diversion, we went to weekly movies at the town's recreation hall. These movies were invariably comedies or musicals to keep up the spirits of the population. In the last months of the war, the films had themes of German history in which sticking it out led to victory. In addition, we occasionally took the train to Munich to see friends. At every successive visit the increasing destruction of the city became more apparent. In contrast, during the two-hour trip from Kochel, villages, churches, farm buildings, and all other structures looked as peaceful as ever. But as time progressed, even hiking became restricted by the lack of food, though in the summer we found on the mountain slopes wild strawberries and other fruit that would normally have been harvested. In the last winter of the war we received skis for some mysterious reason, most likely from the surplus of winter equipment collected for the army to keep it from freezing in Russia. I began to ski rather than walk to work, and we improved our skills during our ample free time.

During one excursion, cycling along the lovely Walchensee after pushing the bicycles up the Kesselberg, we made an interesting discovery. On the shore of the lake we saw a construction site. We heard from some local people that the National Socialist party headquarters in Munich, the well-known Brown House, was to be evacuated and a new office complex was to take its place here. In the midst of heavy bombing, military disasters in Russia, and the anticipation of the Allied invasion of France, a costly building project was being undertaken for the party. We heard that elaborate interior decorations, special furniture, and palatial accessories were being installed by many artisans. All this was being done for the leadership of a party that exhorted the population to fight a total war and to put up with all hardships to achieve victory.

At some time after my arrival in Kochel, I was made the head of the *Werkschutz*, or plant protection. I do not know who appointed me to this position. As a first lieutenant among the soldiers from the air force and army assigned to the wind tunnels—I believe that there were eighteen of us in all—I was one of the two officers available for such a job. The other was my friend Günther Hermann, who held the same rank in the army as I did in the air force. As assistant to the director, Günther did not have time to perform the extra, somewhat nebulous duties of plant protection. Unlike the civilians, soldiers were at least safe from the draft. But our little group of soldiers could easily have been assigned to fighting units in the last phase of the war. I believe, however, none of us ever seriously considered that possibility. At any rate, I was now supposed to direct the defense of the wind tunnels if it came to that, guard against sabotage, and act as air-raid warden. The last-mentioned job was easy. When the sirens sounded, we stopped the machinery, left the buildings, and walked to the nearby woods. There we sat around until a single tone of the siren signaled that the aircraft—as usual on their way to Munich—had passed Kochel.

One day a strange event occurred. By chance, I was outside when I heard an unusual crackling noise. Looking around, I saw a continuous lightning-like arc discharge between the top of the low wooden building next to ours and the high-voltage transmission lines right above the building's roof. A moment later, the few people working in the building rushed out, some of them scrambling on hands and knees. Next, flames sprang up instantaneously along the length of the building. In spite of our efforts and the help of the Kochel fire department, the structure burned to the ground. The most we could do was to keep the flames from spreading to other buildings, and the wind tunnels were not touched. Fortunately, nobody was hurt during the period in which everybody was exposed to a high electric potential. The occupants reported later that they were stunned by an unknown effect, presumably the potential prior to the discharge, and could barely move.

I was aware at the time that this fire was adding to the suspicions attending our secret installation—and us—in the eyes of the local population. After the cleanup and a cursory investigation of the fire's origin, we discovered that in order to speed up construction, standard safety rules prescribing distances between transmission lines and buildings had been waived. Prior to the fire, an American daylight raid had taken place, and the high-voltage transmission line was destroyed not far from Kochel. Power transmission was stopped instantaneously, and the line shorted out. The subsequent surge of power apparently caused the electric arc that I saw.

Occasionally, American aircraft were shot down or damaged and subsequently abandoned on their way back to Italy, with the airmen parachuting to safety. One such incident took place not far from the wind tunnels. A group of wind-tunnel people set out to find the American crew. I was not among them, but I was told later that after the fliers were located, one wind-tunnel staff member tried to attack an American flier but was restrained by the others in the group. The staff member had lost his wife in the Peenemünde raid, and was acting in a fit of anger.[2]

Normally, American airmen who reached the ground by parachute would be picked up by farmers or others to be brought to the local police. The police in turn delivered them to a prisoner-of-war camp. This policy changed one day when a police officer came cycling to the wind-tunnel administration building to speak to the director and to me in my capacity as air-raid warden. He handed us a directive from the head of the Bavarian police. This document had been issued in the name of Himmler, who by now controlled all regular police units in Germany, in addition to the diverse agencies of the SS. We were admonished to remember that Allied bombings were acts of terror and that airmen who parachuted onto German territory were criminals who might shoot us while nearing the ground, making it unsafe to capture them. They ought to be shot by us in self-defense, with the further positive result of reducing the number who became prisoners of war. The document stated, according to the pleasant elderly Bavarian policeman, that too often the population and local police had delivered captured airmen to prisoner-of-war camps. He read through this directive as if it were an ordinance dealing with the parking of bicycles at the WVA.

After the policeman left, I turned in disgust to Hermann, remarking that we had come to the point where the government wanted to turn us into murderers. Hermann was incensed. He said that he really did not understand me. By this time, he clearly believed that I represented a danger to his institution. In the past, he had occasionally admonished me—and I am sure others—to use the obligatory "Heil Hitler" instead of the standard "Guten

Tag" or "Auf Wiedersehen." Now he took Himmler's side, saying that the air raids were indeed criminal actions. It is curious that an intelligent person like Hermann did not view in a similar light the German air raid on Rotterdam, a city in a neutral country, not to mention the later attacks on London, Coventry, and other British cities.

The summer of 1944 was warm and sunny; serious events seemed remote, the Allied invasion seemed far away, and counting airplanes that did not attack Kochel became a routine game. Of course, Munich was increasingly falling into ruins, the only visible effect of a war that otherwise appeared to be far away. With a little effort one could become quite well informed about the actual happenings of the war, however. The daily report of the military, the *Wehrmachtsbericht,* which had been quite accurate and complete during the earlier phases of the war, still gave geographical locations, accounts of the fighting, names of the German cities that were bombed by the Allies, and many other details that filled out the hidden picture. Everyone quickly learned to interpret these increasingly incomplete reports, and one mastered the technique of reading between the lines. On maps, front lines were "straightened out," a euphemism used to disguise the fact that pockets of land occupied by Germans had been given up. Unexpected names of rivers and towns popped up. These could be found on maps, and the rapidly changing military picture emerged quite clearly. Early in June the expected Allied invasion finally took place. The beaches of Normandy were taken, and the German army had to face a second front in the west. No enemy troops had been encircled or thrown back into the Atlantic as promised, and it became obvious in late summer that the invasion had succeeded on a broad base. The German forces fought desperately to hold their ground, with the Allies exercising near-total control of the air. In his famous book *Mein Kampf,* written during incarceration after the Munich putsch, Hitler had derided the World War I strategy of the Kaiser and his generals of fighting simultaneously in east and west. He now faced an identical situation on a much greater scale, having vastly underrated the power of the Soviet Union.

Although it was punishable to listen to German-language newscasts by the BBC from London, many did so. These broadcasts were by no means universally reliable, but it was obvious even from our own daily bulletins that the Allies were moving across France toward the German border. American transmitters were set up in France in the wake of the advancing Allied army.

Another major event happened shortly after the invasion of France. Against all expectations, an attempt to assassinate Hitler took place on July 20 in his East Prussian headquarters. The first reports were garbled, and I believed for a while—in common with most Germans—that Hitler was indeed dead. The military and civilian conspirators in Berlin and elsewhere also initially believed in the success of the attempt. But as the event unfolded, all of us could soon hear Hitler's somewhat weakened but otherwise unmistakable voice on the radio. Aside from official reports of arrests, rumors of an unprecedented frenzy of revenge and retaliation by government and party soon reached us. Killings, disappearances, and trials were mentioned. Although a large city like Berlin was constantly buzzing with stories, and although the city had remained a place of some openness, the conspirators had kept their plans secret. While one or two of the civilian members of the revolt—which had been jointly carried out by high officers in the army and politicians covering all parties of the Weimar Republic—were known to my father and had visited our house, he had no inkling of their plans.

All members of the armed services, including those disguised as civilians at the WVA, now faced a decision as a result of the attempted assassination. Soon after the upheaval in July, the government decreed new rules of eligibility for membership in the Nazi party. I myself had been safe from solicitations to join the party or any of its organizations since November 1938, when I was drafted. According to a law passed in the early days of the Weimar Republic and following the terms of the Treaty of Versailles, the 100,000 members of the Reichswehr, the permitted rump German army, were not permitted to join political parties. Strangely, that law was retained after 1933, with only the National Socialist party remaining. The highest Nazi official at the WVA, a draftsman whom I knew slightly, appeared in my office shortly after July 20. The Führer, he pointed out, had ordered an end to the shameful fact that officers and men of the armed forces could not join the party, and would I please apply for membership. I thought the proposal ludicrous, lost my temper, and blurted something to the effect that I would have to be crazy to join the party in July 1944.

The man disappeared, but soon the effects of this and, I believe, other happenings caused a frightening development for me. Although Rudolf Hermann was certainly critical of my general attitude and that of others at the wind tunnels, I do not believe that he went as far as reporting me or those who held similar views to the Gestapo, the secret police run by the SS. It seemed that he still wished to see us primarily as engineers and scientists who after all did their jobs properly and committed no overt antigovernment acts. Then again, he had certainly done something about Siegfried Erdmann, who had yet to appear at Kochel after his path-breaking experiments in achieving hypersonic flow. All of us were careful in Hermann's presence.

One day I received orders to report to the Gestapo headquarters in Munich, and this trip turned into a powerful experience in my life. I took the train to Munich and went to the headquarters of an organization whose dangerous activities were by that time known by most people. Any contact with this organization was dreaded, and I was extremely scared. After identifying myself with my trusty Soldbuch, I was directed to a room in an upper story. The building in Dietlindenstrasse was still untouched by the war; it seemed more like an apartment house than a police station. The place looked bare; no personal touch was visible. A stale smell pervaded the grayish corridors. I saw no uniformed person except the guard at the door, and the people I spoke with wore somewhat shabby suits. I am certain in retrospect of the Kafkaesque features of this visit, yet I am uncertain whether this impression was caused by reality or my own apprehensions. I was interviewed by two men; one sat behind a desk and one leaned against it. I say *interviewed* because *interrogated* would be too strong. No specific questions were asked concerning Erdmann's Denkschrift or about my arrest in Pasewalk, my seeming problem with Rudolf Hermann, or several other events that I had been thinking about since receiving the order to appear. I do not recall that we talked about anything other than the wind tunnels, whose existence was of course no secret to the Gestapo.

In fact, by late 1944 Himmler had achieved control of all V-weapons and antiaircraft-missile development and production. Possibly they just wanted to find out whether I could be used. After some time—short or long?—I was told that I could return to Kochel. Outside on the street I felt relieved and breathed freely. Shortly after this visit I was relieved of my job as head of the Werkschutz, but nothing else happened. I was reinstated during the last weeks of the war early in 1945. Consequently I had a good opportunity to affect the fate of the WVA during the days prior to the arrival of the American army in Kochel.

As an aside, I must add a more personal note. Since the beginning of the offensive against the Soviet Union, I always had the feeling that it would be exceedingly difficult to escape unscathed from the cataclysmic events enveloping most of Europe. This feeling was alleviated somewhat by the move to Peenemünde, quite aside from the relief of not being shot at. The move to Kochel and the many breathing spells of a "normal" life, such as hikes in the mountains, did much to relieve pessimism. The overall events appeared to me to be crushing, however; I was unable to imagine what the end might be like.

Although I knew that Siegfried Erdmann needed to remain up north to finish the closing of the facility, he did not show up in Kochel for some time after my own arrival. My recollection—faulty, as we shall see—tells me that one day I received in some fashion the results of his novel high-Mach-number experiments. Data records, photographs, and handwritten notes by Erdmann were handed to me in a folder, with a request that I put the information together in a systematic sequence. I enjoyed the work, since I was fascinated by the experiments, and I returned my compilation to Kurzweg. A short time after this episode I was made head of the basic research group. This was, of course, Erdmann's position, and the fact that I had become his successor indicated to me that he would not be coming to Kochel at all. I was worried about his fate, since I remembered Erdmann's memorandum on the move to Bavaria and his obvious clash with Hermann. But I was not to see him again for a long time.

In 1984 I was working with a friend at the Technical University in Berlin on experiments carried out jointly with my laboratory at Yale. There I discovered two facts that demonstrated that Erdmann was alive and well. I knew from Peenemünde that Erdmann had the degree of Diplom Ingenieur. He had completed his studies in September 1939 at the very university that I was visiting. Now I discovered that in 1951 he had received a doctoral degree from the Technical University of Aachen. My friend in Berlin had used results of Erdmann's dissertation for our joint paper. In addition, I was told that Erdmann was now a professor at the Technical University of Delft in the Netherlands. There he had built a major aerospace science institute, including supersonic wind tunnels. In spite of his frequent travel to the United States, our paths had never crossed, and I had never heard from him, since his field was different from mine. What had happened?

In view of the happy ending of this story, I was determined to find out the details of the Erdmann saga. In October 1991, I traveled to Pijnacker near Delft, where Erdmann—now retired—lives with his wife. In two days of fascinating discussions about Peenemünde, I learned much about the time prior to my own arrival at the Baltic. But first I found out that Erdmann had indeed shown up in Kochel. He himself, in fact, had handed me the notes on his experiments during a brief meeting in July 1944, a meeting that I had forgotten.

The memorandum that had caused Erdmann's problems with Hermann was recently found by Erdmann in an archive, and he sent me a copy. Following in part a wish of Wernher von Braun, Erdmann had suggested splitting the aerodynamics activities. The local top authorities arranged to retain the large Peenemünde vacuum vessel in order to keep one of the wind tunnels under their control. This fact showed that in the tug-of-war between Hermann and von Braun, the latter and his staff opted for Erdmann's plans. In retrospect, this proposal seems like a reasonable plan considering the

The Photographs

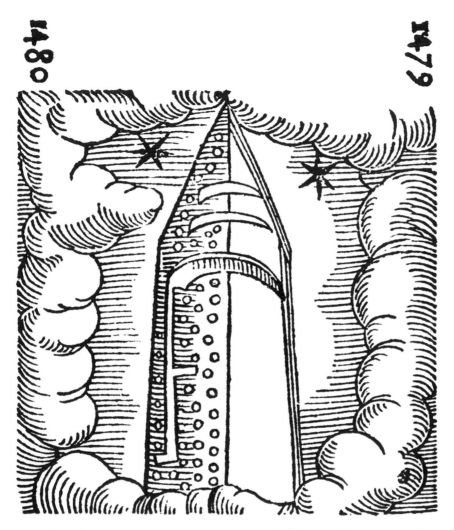

A woodcut of a spaceship dating from 1480, published by Conrad Lycosthenes of Basel in 1557. (DMB)

A private rocket testing ground near Berlin in 1930. On the left, Rudolf Nebel; on the right, eighteen-year-old Wernher von Braun, a high school student. (DMB)

The A3—forerunner of the A4—on a test stand in Kummersdorf, an early rocket testing ground of the army, also near Berlin. Taken from a secret army report of 1937. The tail end is not yet shrouded. (DMB)

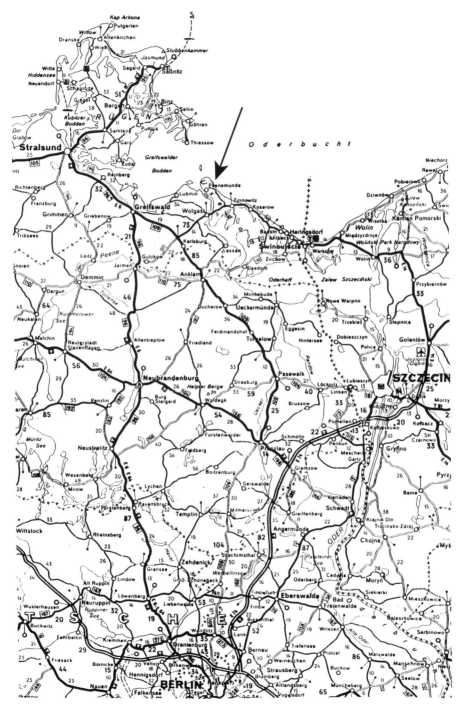

A map of part of northern Germany shows the location of the new army Research Station, Peenemünde on Usedom, an island in the Baltic Sea separated by a narrow bay from the mainland. The exact location is marked with an arrow. (PW)

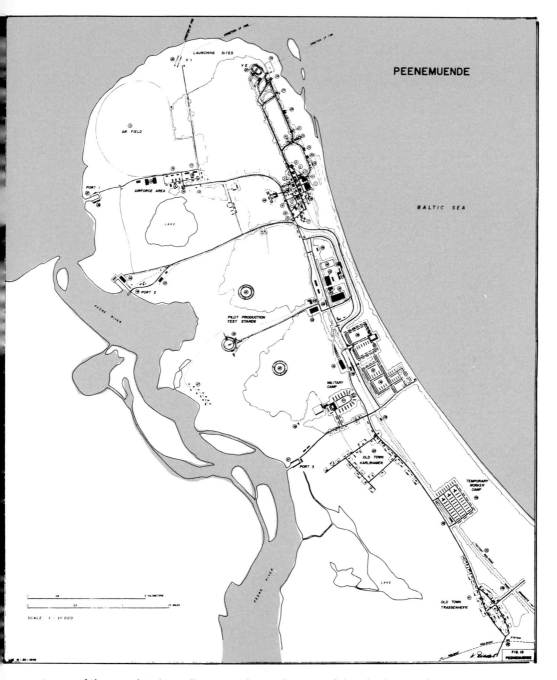

A map of the completed installation. At the northern tip of the island, an air force laboratory—including a small airport—was the site of the development of the flying bomb later called the V1. The army power plant is at the edge of the water nearest the mainland. (DMB)

Gatehouse with guards leading to the army installation. (DMB)

The A4 rocket prepared for the fourth flight test, in June 1942. The logo contains the only reference to a moon shot that I saw on a Peenemünde photograph. (DMB)

An A4 rocket explodes shortly after leaving the main test stand (November 1942). (DMB)

An explosion at the same site about 2.5 seconds after ignition (January 1943). (DMB)

The rocket engine of the A4. Fuel and oxygen are injected at the top to ignite in the combustion chamber. The exhaust gases reach supersonic speed at the exit after passing through the converging-diverging nozzle. (DMB)

A group of senior officers watching a firing. On the left is General Walter Dornberger. Rudolf Hermann, the director of the Aerodynamics Institute, is in the foreground, wearing a light jacket. In the middle, in a dark suit, is Wernher von Braun. (DMB)

A successful A4 firing late in 1943. (DMB)

A crater 40 meters (130 feet) in diameter and 15 meters (50 feet) deep is caused by the impact of an A4 without an explosive warhead. (DMB)

Peenemünde visit of the armament minister, Fritz Todt (second from left in profile). General Walter Dornberger is in the middle; General Friedrich Olbricht is second from right. Olbricht, active in the attempt to depose Hitler, was shot at army headquarters on July 20, 1944, the day of the failed assassination. The civilian in the background on the left is Heinrich Lübke, an official of a construction company active at Peenemünde, who after the war was a president of the German Federal Republic. (DMB)

The main building of the Aerodynamics Institute housing the two large wind tunnels. The protruding part of the building contains the vacuum tank that was left there when the institute was moved to Bavaria. (PW)

Entry hall of the Aerodynamics
Institute. (PW)

The test section of one of the 40-by-40-centimeter supersonic wind tunnels. During an experiment, the air blows from right to left. A model to be tested is mounted on a rod. (PW)

The heavy glass walls of a nozzle used to produce a supersonic Mach number shattered on one side of the test section, while I had my nose pressed to the other side to view the experiment. Flow in the nozzle would be from right to left; note the expanding top and bottom walls required to reach supersonic airspeeds. (PW)

An electromechanical balance with a model of an artillery shell mounted on a strut. When installed in the test section, the balance is used to measure aerodynamic forces like drag or lift at various angles of the model with respect to the airflow. (PW)

One arm of an optical system riding on top of the test section. During the experiments, schlieren photographs (see chapter 3, note 4) are taken of the flow. (PW)

Schlieren picture of a model of the A4 at supersonic speed. Note the shock waves originating at the tip of the model. (PW)

A model of the antiaircraft rocket Wasserfall in the test section. The model can move freely in the vertical plane. (PW)

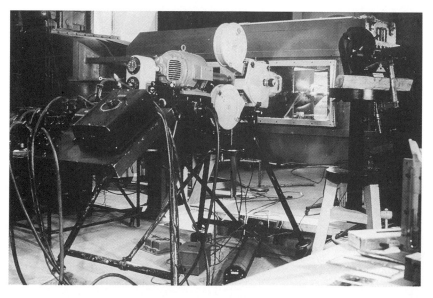

A high-speed camera set up to follow the motion of a model such as the Wasserfall. (PW)

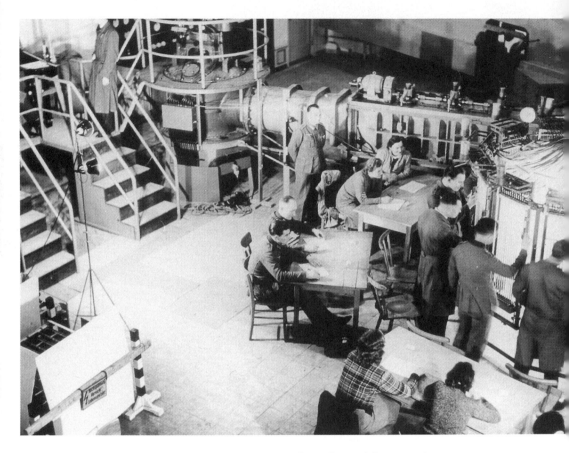

An experiment to measure the pressure on the surface of a model at many locations. Small holes in the surface were attached to tiny tubes that led to mercury manometers read by many helpers. (PW)

An exhibit of models of different missiles, artillery shells, and bombs that were tested over the years in the Peenemünde supersonic wind tunnels. The two drawings on the left depict the A4, later called V2. (PW)

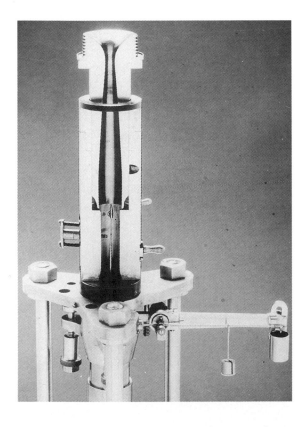

For comparison with the advanced supersonic wind tunnels of the 1930s and 1940s at Peenemünde, here is the earliest supersonic tunnel that I could find. Toward the end of World War I, this facility operated at the National Physical Laboratory (NPL) in England. At the top, a steel tube (not shown) supplied air at a pressure of 50 pounds per square inch. Beyond the narrow section at which the speed of sound is attained, the expanding nozzle, just like the nozzles of Peenemünde, produced a supersonic flow in which the aerodynamic drag of a model of an artillery shell was measured. The model is attached to a mechanical balance, seen on the right. The test section's diameter is roughly that of a finger, and the model is as small as a pencil tip. This tiny facility is cut open for viewing and can be seen in the NPL museum in Teddington, England. (NPL)

Aerial photo of the Peenemünde settlement after the first—and heaviest—air raid, on the night of August 17–18, 1943. (DMB)

The A4 production facility at Peenemünde after the air raid. (DMB)

View of Kochel, Bavaria, the small town to which the Aerodynamics Institute was moved after the first air raid. (PW)

A bird's-eye view of the Walchensee (pictured above) and the Kochelsee, which are connected by pipes to operate Germany's largest hydroelectric power plant, located at the bottom. (PW)

Scenes of the construction at
Kochel in preparation for the arrival
of the wind tunnels from
Peenemünde. (PW)

A schlieren photograph of a model of the A4b, the winged long-distance version of the A4. (PW)

The main transmission line from the Kochel power plant was destroyed in an air raid north of Kochel. The short circuit caused an arc discharge that burned down a storage building next to the wind tunnels. I am in the foreground; behind me is a local fireman. (PW)

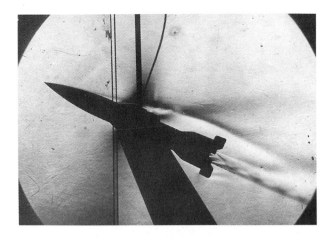

Simulation of the rocket jet with a wind-tunnel model of the Wasserfall. (PW)

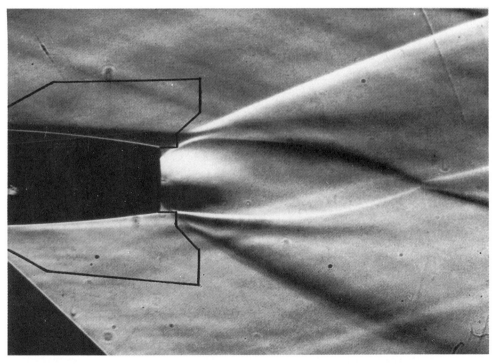

The tail end of a model of the A4 in the test section of the wind tunnel operating at a Mach number of 3.2. The rocket jet is simulated by a high-pressure jet of air. The expansion of the jet must be known so that damage to the fins can be avoided. The schlieren picture shows the structure of the jet. (PW)

The first photo ever taken of hypersonic flow at a Mach number close to 9. The experiments were the final ones carried out by Siegfried Erdmann at Peenemünde before the last wind tunnel was moved to Kochel. (PW)

A model of a projected hypersonic wind tunnel with a one-square-meter test section. Kochel had been selected as the site for this project long before the air raid because there would be sufficient power to run the huge facility, which was never built. (PW)

Preparations for the field use of the A4 included trials of a special trailer that could transport the missile and then serve as a firing platform. (DMB)

The first battery of artillerists practices in the summer of 1943. (DMB)

A practice shot at a hidden launching site. (DMB)

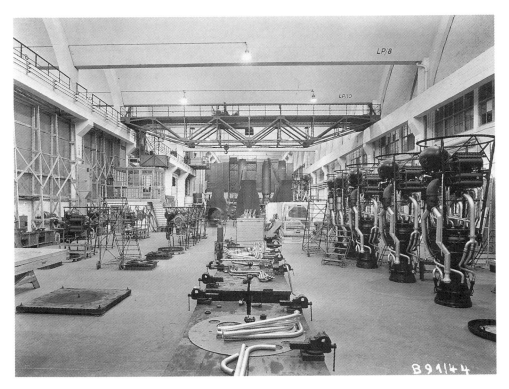

The rebuilt A4 assembly building at Peenemünde in 1944. (DMB)

Transport of an A4b to a test stand at Peenemünde in 1945. (DMB)

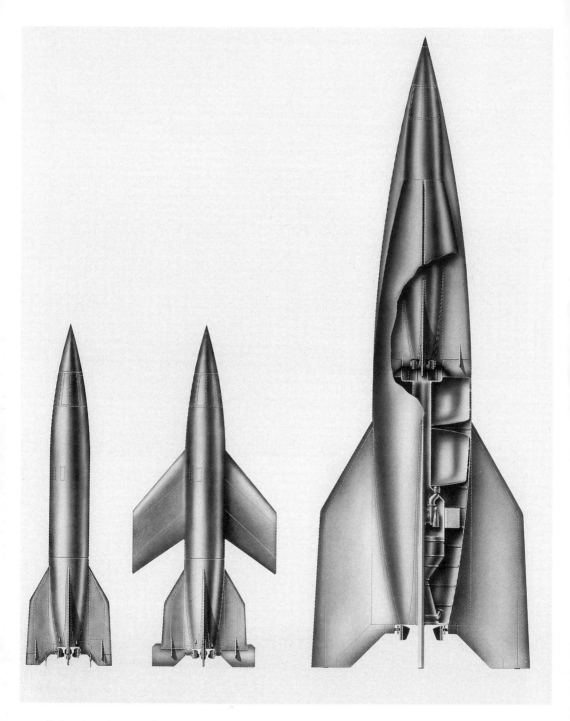

A drawing showing three Peenemünde projects. From left: A4 (V2), A4b, A9/A10. The development of the tall two-stage missile was never seriously started. (DMB)

Festivities at Peenemünde at the time General Dornberger and von Braun received the civilian equivalent of the Ritterkreuz (Knight's Cross). (DMB)

A 1992 view of the Kohnstein, in whose underground tunnels—the Mittelwerk—the A4 and other weapons were mass-produced. At left is the museum memorializing the concentration camp Dora. The opposite side of the mountain is an open-pit mine. (PW)

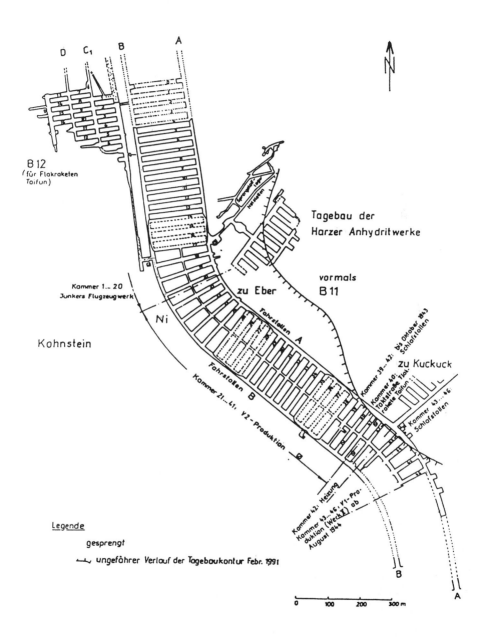

A plan of the Mittelwerk. The V2 was assembled in tunnel B, between the point marked Ni and the lower exit. The dashed lines show the areas that were blown up by the Soviet army after the war. The tunnels are 1.8 kilometers long (1.1 mile), and the distance between the main tunnels is 180 meters (600 feet). (Dora)

One of the four main entries to the Mittelwerk. The two main tunnels were equipped with tracks. (Dora)

V2 assembly; tanks and insulation are mounted. (Dora)

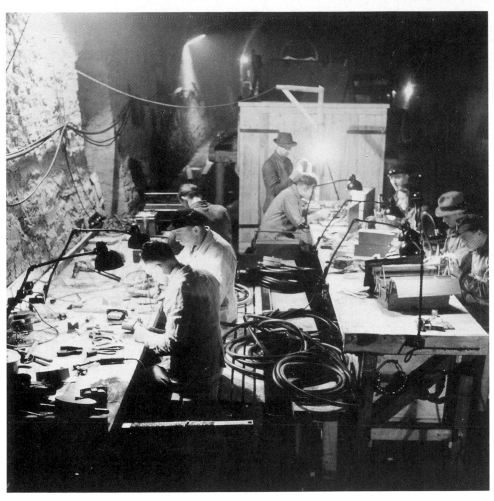

A laboratory in the Mittelwerk. This is said to be the only existing photograph taken inside the plant during the war. The workers may not be concentration camp inmates. (DMB)

Machinery in one of the cross tunnels in the Mittelwerk. (Dora)

The main entrances to the Mittelwerk were blown up by the Soviets before they left the site. This entry was dug by the staff of the memorial site. (PW)

The remnants of a V2 rocket motor recently photographed inside the closed Mittelwerk. (Dora)

The defunct coal-dust power plant of Peenemünde, located on the bay side of Usedom, that now houses a small museum of Peenemünde history. A Soviet MIG fighter aircraft is exhibited. (PW)

progress of the war and the frantic efforts to develop further missiles, such as the long-range successors of the A4. Again, however, I am happy that these plans never came to pass, because in Kochel I was far from Peenemünde, which experienced truly difficult times toward the end of the war.

As was to be expected, Hermann strongly opposed the dismemberment of his institute. For one thing, he would have lost control of the part of the operation that involved new projects, in particular the work on an intercontinental missile. In those days, when only projects of extreme urgency could secure equipment and labor, attention might have been shifted to the Peenemünde rump operation associated with von Braun. Characteristically, the postwar report by Hermann and others on the achievements of the Aerodynamics Institute, including the design of the Kochel hypersonic wind tunnel, simply ignores one of the most interesting experiments, Erdmann's pioneering high-Mach-number work. In fact, Erdmann, who also did much of the groundwork required to perform wind-tunnel experiments at supersonic speeds, is not even mentioned. It is difficult to see how anyone could still have believed in the fall of 1944 that resisting a reorganization of the Aerodynamics Institute could affect the course of the war. It is even more difficult to see how anyone could have wished to put Erdmann at extreme personal risk. And yet that is what Hermann proceeded to do.

After completing the remaining work at Peenemünde, Erdmann traveled to Kochel and was immediately confined to his quarters by the director. During a single brief appearance at the wind tunnels, however—the visit that I had forgotten—he handed me the material on his hypersonic flow experiments. (I hope that readers of these recollections will not assume that similar failures of memory compromise too much of what I have written.) At any rate, Erdmann's arrest explains why I had been asked to take on responsibility for the basic research group.

Erdmann further told me in 1991 that Hermann had contacted the army headquarters in Berlin, circumventing the Peenemünde command. Hermann accused Erdmann of negligence in the handling of secret documents. This accusation was taken seriously, and a Major General von Junck traveled from Berlin to Kochel to investigate. A meeting was convened at the WVA, where the director and top staff—including Kurzweg and Eber—together with the general were to decide Erdmann's fate. The accused was asked to retract his by now famous Denkschrift. He refused to do so, and on September 7, the day after the meeting, he was drafted into the army and ordered to join a heavy artillery unit stationed in Landsberg, a city close to Munich. After he left Kochel, Hermann convened a meeting of the entire staff of the WVA. He read a part of the Denkschrift, and he added that Erdmann had acted to advance his own career, prohibiting further contact with him. In retrospect, I am sorry that I not in Kochel during the time of this meeting, and nobody ever mentioned it to me.

At the same time, an army court-martial was instituted, and in addition Erdmann was ordered to report to an office of the SS in Berlin. He was fortunate that the court proceedings dragged and that an SS officer in Berlin found the case unjust and tried to help him. He was never assigned to a unit involved in combat but rather spent much of the rest of the war shuttling among the military court, the SS office in Berlin, and the barracks in Landsberg. I am sure that the difficulty of travel during the last months aided his delaying tactics. He survived by perseverance and luck. Miraculously close to the end of the war, he was again assigned to an aeronautics research establishment at Brunswick.

After the war Erdmann met with Wernher von Braun at Wimbledon in England. By now, it had developed that von Braun and his group might move to the United States, and von Braun asked him to join them. Erdmann decided against this and went instead to the Netherlands and then to Sweden. Late in 1954 he accepted an offer from Delft, where he built up the institute I mentioned before. Even during the last year of the war I would not have thought that Hermann would have gone to such lengths to persecute one of his most able scientific staff members.

Late in 1944 my group was sidetracked from our self-generated basic research. It was a time when the last resources of Peenemünde and similar laboratories were being directed toward finding a way of defending the cities from the increasing air raids, which were often carried out in broad daylight. In contrast with the area bombing of cities at night, the raids could be directed specifically against transportation centers and other targets, such as factories. These attacks were usually carried out by the American air force. I heard that Lehnert's group was working on projects in the wind tunnels and elsewhere, all directed to the development of antiaircraft rockets. These devices did not follow the ambitious design of a supersonic missile like the Wasserfall. The new weapons looked more like kites or small airplanes, all moving at speeds lower than the speed of sound. They carried colorful and inappropriate names like *Schmetterling* (butterfly) or the Wagnerian *Rheintochter* (daughter of the Rhine). There was also a parallel development of tiny one-man rocket airplanes, which were actually completed and used in the field. We did not even hear of that at the time, however.

Our group was also called on to do some aerodynamics work along such lines. A solid-fuel rocket called *Taifun* (typhoon), a small cylinder roughly as tall as a man, was to be fired supersonically in clusters at a formation of attacking aircraft. Because it did not have a guidance system, it had to be pointed directly at its target. Aerodynamically speaking, all it required was stable flight, just like that of an arrow. We quickly found the proper tail-fin configuration to keep the flying pipe from tumbling, and the results were reported in November 1944. As far as I know, nothing ever became of this device. At any rate, this work was the last requested from above (army or air force?) that I was involved with. Otherwise, the last months of the war caused little change in the daily research routine at the wind tunnels. We resumed the experiments on sphere drag, tried without success to make aerodynamics measurements at exactly the speed of sound, and tinkered with other problems.

Along with the decreasing workload, an increase in the efforts to sustain our daily lives became necessary. At a certain point, those of us without children were moved to the Hotel Stöger, located opposite the small railroad terminal. The central heating had to be curtailed because of the lack of coal; however, small woodstoves installed in every room were supplemented by electric space heaters. We had to cut our own firewood. The town government provided tools and assigned a stand of fir trees to cut down. To make life tough for us rank amateurs in the lumbering business, they selected a steep slope. I had brought at least some experience with me from my service in the Arbeitsdienst, when I had worked in the woods. After cutting a tree down and doing all the work from transporting to splitting the logs, we piled the green wood on top of the stove to dry it out. Fortunately, we didn't burn down the hotel.

Early in the year the town had also assigned land to grow vegetables. I planted seed potatoes, which did poorly in the rocky soil at such relatively high altitude. I did harvest nice new potatoes, but the total weight of my new potatoes was just equal to that of my seed potatoes. Kurzweg and I had bet which of us would harvest more tomatoes. I won because my plot produced a single small tomato and his had none. During the last winter of 1944–1945, we tried everything to find more food. My wife occasionally got some blood from the butcher, and if a turnip came our way we were happy. All this even though the local population was well fed and continued to barter with us.

My last trip to Berlin before the end of the war was undertaken in connection with top-secret orders to attend a meeting on September 23, 1944, in Kölpiensee, a small lake next to Peenemünde whose name was used to disguise our appearance at the laboratory. This technical meeting of a relatively high-level group was convened to deal with the progress, or lack of it, of the Wasserfall. We had long finished the work on Rudder 21, our best solution to the problem, and I reported to the group on the status of the aerodynamics of the missile. It became clear during the discussion that the completion of this ambitious project was still far in the future in spite of some successful test launches. The minutes of the meeting were duly re-corded, and years later I found a top-secret document—including a tran-script of my own remarks—in the archives of the Deutsches Museum in Munich.[3]

I traveled in uniform, a much safer way to get around and to obtain transportation. On the way back I took my time and spent a few days in Berlin visiting my father.[4] His house had a new roof and new glass in the windows, and it was in reasonable shape. But by now much of the city was in ruins. A semblance of normal life was nonetheless still possible. The food distribution system, no matter how small the supply at this time, still func-tioned. The subways worked well, although—in contrast with London's—the tunnels were just below the surface. Heavy bombs often penetrated the tunnels, and obviously the stations could not be used as bomb shelters. But the track could be quickly fixed without closing the holes in the ceiling. My father and the other Berliners that I spoke with expected the Soviet army to be the first to arrive in the capital, although in October 1944 the troops were still relatively far away. I returned to Bavaria by train without any particular problems, not realizing that I would never see my father again.

As before, the Bavarian countryside had a calming effect on me after the many discussions in Berlin with friends of my father, who knew much more about the war than I did. The end of the war was approaching rapidly, but a few important incidents happened before the American troops entered Kochel.

EIGHT

The End of the War

I VIVIDLY REMEMBER CHRISTMAS EVE 1944. WE WENT to midnight mass at the beautiful church in Kochel, joining a group of onlookers at the back. The service represented a world apart from our own, but there was some comfort in its reminder of continuity with the past. As I recall, the priest's brief sermon did not allude to the war, the many young men of Kochel who had died in it, or what might lie ahead for all of us in the coming year.

It was clear that 1945 must see the end of the war in Europe, although one did not know exactly when. I believe nearly everybody knew by then that Germany had lost the war. Rumors still circulated, however, about wonder weapons that were to ensure victory at the last minute. We at the wind tunnels were obviously the last people who might expect any such weapons beyond the V1 and V2, and even those two were hardly mentioned in the daily military reports. What else could exist? I do not recall a single discussion of the possibility of an atom bomb. The British raid on the heavy-water plant in Norway and Werner Heisenberg's efforts to produce a nuclear reactor were unknown to us. Moreover, none of us had been aware of the 1939 paper by Otto Hahn, published in a German scientific journal with a worldwide distribution, discussing uranium fission, the scientific basis for a nuclear bomb. Heisenberg's family lived at Urfeld, a town on the pass between the Walchensee and the Kochelsee. An excellent young drafts-woman in my group had left us to become an au pair with Mrs. Heisenberg, helping with the extended family. She never said anything about Heisenberg's work and probably knew nothing about it anyway.

An increasing number of us began to think about the future. We expected occupation by the American army, and we considered what might be done with the wind tunnels. Nobody ever suggested blowing up our laboratories. More and more of the scientific people and others began quiet discussions about ways to preserve the equipment. Since the wind tunnels and the large machinery installed at Kochel could not be hidden in the woods, we considered preserving special equipment, such as instrumentation. For example, we planned to take apart the optical equipment—including the beautiful Zeiss interferometer, of which no other example existed—seal the components in watertight cases, and bury them in the woods. In addition, the original drawings, calculations, and reports in which the results of the aerodynamics experiments were recorded had to be saved. While we were planning for the last days of the war, grim scenarios of what might happen to the institute came to mind, such as destruction by retreating troops or by special German commando units assigned to blow up bridges, power stations, and the like; looting by soldiers of whatever army; and, most important, destruction by local zealots of the Nazi party who had some idea of what was going on at the WVA. Those who did not actively support the efforts at preservation—efforts that could not be kept totally secret—at least kept quiet. I do not know what the director thought about all this.

To put these plans and the events to be described here into context, I note that in January 1945 the Soviet army entered East Prussia and Czechoslovakia. At the end of the month, the part of Germany east of the Oder River (the current boundary with Poland) was in Soviet hands. Next, the German army—now under the command of Himmler—tried to defend the Oder line, by now a hopeless undertaking. Nothing but defeats followed, culminating in the battle for Berlin. On March 7 the Allies on the western front crossed the Rhine, and from then on, no serious resistance was encountered. The unconditional surrender of Germany was signed at Rheims on May 7 and ratified in Berlin on the following day.

The original reports in which the research of the Aerodynamics Institute was preserved dated from the founding of the institute in Peenemünde in 1937 to the final work at Kochel. These reports were all labeled as the Archive 66 series of the Army Research Institute, later of the WVA. All originals were typed, and the figures were drawn in ink on vellum, a tough, transparent paper. Original photographs were glued down on the same paper, and typed captions were added. For future research, we wanted to safeguard this series of papers, which included about two hundred documents of varying lengths. Surely this was the most comprehensive body of work on experimental and to some extent theoretical supersonic aerodynamics existing anywhere, as we now suspected.

We found out that unfortunately the originals of all documents from the pre-WVA period—the longest period of active research and development—

were not at Kochel. We did not have even a complete set of reproductions of the originals. It turned out that the so-called Archive 66 series, like most Peenemünde documents, had been stored for safekeeping in the underground factory mentioned in chapter 6. The Kohnstein, a long hill about 100 meters (330 feet) high with existing mine tunnels, had been selected before the first air raid on Peenemünde as the site for the mass production of the V1 and the V2 and other military projects. Production had started in September 1943. The underground factory, called the Mittelwerk, was located near the small town of Nordhausen on the eastern side of the Harz Mountains, an area that was later part of the German Democratic Republic.[1] It was decided with Rudolf Hermann's blessing that my friend Günther Hermann and I should travel to the Mittelwerk to retrieve the documents from the clutches of unknown authorities.

The director had given us travel orders that implicitly required us to take the shortest route between Kochel and Nordhausen. But such orders issued by a small, unknown research station would be of doubtful value if we were stopped. The distance from Kochel to the Mittelwerk and slightly beyond to Magdeburg is about 650 kilometers (400 miles), and a 1,300-kilometer round-trip made at this terminal stage of the war was an ambitious undertaking indeed.

Günther and I gathered as much food as possible, filled a number of gasoline cans, depleting the wva supply, took our hidden uniforms, and started out around Easter 1945 in our battered Opel Olympia painted army gray. This trip turned out to be a true adventure. Once our supplies ran out, problems in obtaining gasoline and food arose. Bomb craters pitted the roads, and a nighttime blackout complicated driving. Primarily, however, we worried that our very appearance as two healthy, low-ranking officers would tempt those of superior rank who commanded active units to draft us into the fighting. Our arrival at the Mittelwerk would be unexpected. How could we prove our right to the archives? We even wondered whether our request could be viewed as a defeatist act, since it would be difficult to guarantee the security of the documents while returning to Kochel in the dilapidated car.

I am amazed now at how little I worried—if I did at all—about these well-understood difficulties. We would get a good view of conditions in the middle of Germany, far from our peaceful Kochel. I remember that I was also eager to find out how literally thousands of V2 missiles could be built in underground caverns in a country whose infrastructure was close to total collapse.

Since we had no prescribed schedule, I hoped to go briefly to Berlin to see my family. (Today it takes about three hours to drive from the Mittelwerk to Berlin.) The final fight for my hometown had not yet started. Günther, on the other hand, had in-laws in Magdeburg, northeast of Nordhausen and much closer than Berlin. His relatives owned a flour mill and so were well provided with food by barter, and he hoped to bring some of their abundance to Kochel to share with his wife.

Our early departure was uneventful. Shortly after leaving Kochel, we put on our uniforms. We spelled each other and drove all day. Strangely, there was little traffic on the north-south routes. Supplies for the armies were carried by train, intermittently or not at all. I recalled Hitler's remark, upon dedicating a new autobahn, that during future wars cattle would graze where railroad tracks used to be. To our surprise, the trip continued to be quite uneventful. At night, we parked off the road and viewed air raids from a distance, much like far-off fireworks. We arrived in the vicinity of Nordhausen on the second day.

Before I describe our search for the underground factory, it is important to note how much had changed since we left Peenemünde. While we labored at Kochel, Himmler's SS had in 1944 wrested total control of rocket development from the Army Weapons Office (Heereswaffenamt). The SS general Dr. Hans Kammler, an engineer, became the supreme commander of all efforts in rocketry, and in addition to the regular staff he marshaled enslaved foreign scientists and engineers and groups of specialists and laborers drawn from concentration camps to speed up production.[2] Another transformation of the army research laboratory took place when extensive daylight bombing of Peenemünde occurred. The Soviets approached along the Baltic, and the whole place was abandoned in steps. Many of the technicians, construction workers, and railroad operators came from villages and towns on Usedom and nearby on the mainland. They could simply go home when the final crisis emerged. Others joined the laboratories and production facilities in the Mittelwerk. People and equipment were moved from Peenemünde to the Mittelwerk by train and caravans of trucks, and housing was arranged in the area of Nordhausen. Some of the Peenemünde staff had already left to help the army's frontline artillerists with the complicated procedure used to fire the V2 on enemy targets. (It seemed amazing, considering the difficulties of production, transportation, and final setup for firing near the front, that a large fraction of the missiles hit such areas as the south of England and Antwerp and other places to which the Allies had advanced.) Wernher von Braun occasionally traveled to the front units, but he was often at the Mittelwerk with his crew. Many problems still plagued the missile, and technical changes were made up to the last minutes of the war. In addition, work still proceeded in Peenemünde on the long-range, winged version of the V2.

But back to our trip. It was not easy to find the super-secret Mittelwerk, and we drove around for a while, trying in a noncommittal way to obtain directions. Finally, we located von Braun, who—if I remember correctly—worked in requisitioned buildings near the underground plant. I knew him well by now, and I felt that we could talk with him openly about our plans and possible problems. As noted before, this gifted and complex man was not ideologically inclined, and certainly he was no Nazi. We shared with him our fear of making a wrong move, since he obviously knew with whom we had to deal once we got inside the plant.

Von Braun provided us with the missing link of our plan: orders to enter the Mittelwerk to retrieve the wind-tunnel documents. Next he arranged for passes in our names, which were signed by SS General Kammler or someone in his office. The passes, covered with all kinds of stamps, looked impressive, and they directed us to transport top-secret documents to the Alps. We had heard that a redoubt was planned in the Alps for a last defense of the regime. It was said that this imaginary fortress was being equipped to support troops to hold out for a long time. In fact, rumors to this effect persisted until the end of the war. Commanders we might encounter on our way south were ordered not to requisition us for their active units but to support our secret and extremely important mission, effectively removing our primary worry.

Thus provided with Hermann's letter to retrieve the archive series and with von Braun's and Kammler's office's orders to enter the Mittelwerk and travel back to Bavaria, we drove the short distance to the underground plant. We located the entrance at the base of the rather unimpressive hill. Railroad tracks on one side of the hill led to the two parallel assembly lines. Some of the massive portals of the old mine looked exactly like those of many nineteenth-century railroad tunnels found all over Germany. The portals admitted materials and prefabricated parts from all over the country, which were assembled inside to produce the V2 missiles. The place was a beehive of activity. Stoppages, we were told later, occurred only when the delivery of parts was interrupted because of the enormous transportation difficulties at that stage of the war. At the side of the main doors, a small entrance was guarded by a soldier. We walked up, showed our passes, and entered the dimly lit interior without delay. The tunnel that we entered seemed particularly dark because we had come in from brilliant sunshine, part of the perfect weather that prevailed for much of the spring and summer of 1945. With German bureaucracy functioning efficiently to the last moment, we were informed that the reports we were after were stored at a designated location.

The tunnels blasted out of the Kohnstein had walls of bare, scalloped rock in most places. Where the rock was loose, concrete had been applied. Workers walked along the tunnels to places where chunks of rock had fallen off the ceiling and applied whitewash to the bare spots. Those parts of the ceiling that were not white could easily be identified to determine whether shoring-up was required. We heard that a few days earlier a large rock had come loose, burying workers who operated a machine tool. Light fixtures were attached to the machine tools in the main tunnels and in the smaller tunnels that connected them. The rockets were assembled on rail carriages that were moved from station to station. (We saw only a part of the V2 assembly line.) Spaces resembling offices—with desks and drafting tables—had been constructed in niches blasted from the rock. At one point in our search for the place where our documents were stored, we ran into a person we had both known at Peenemünde. He became our informant, and much of what we learned about the operation we heard from him.

We noted three different groups of men at work, each in distinctive attire. (On the production line no women were in sight.) First there were Germans from Peenemünde and other laboratories. They ranged from supervisory personnel in engineering and science to master technicians and craftsmen of the many disciplines that were required to put together a radically new device like the V2. These people were, of course, free to move at will, and they were quartered with their families in nearby villages. A few members of this team who had academic backgrounds later became part of the group that fled with von Braun and Dornberger to Bavaria at the end of the war. We were told that the technical director of the mass production of the V2 was Arthur Rudolph, a civilian who had led the pilot production of the missile at Peenemünde. I vaguely remember that I had met him in a small group one evening prior to my move to Kochel, and he had impressed me as forceful and decisive. At the Baltic he had worked on the complex transition of the A4 from an experimental missile to a device that could be produced in large numbers. How much he or his superiors in the SS were responsible for the harsh methods employed to achieve success, I had no way of knowing.

During assembly, the gyroscopes had to be installed and checked. For this and other testing purposes, the V2 had to be raised to a vertical, or firing, position. A huge cavern had been blasted out of the rock, where a powerful overhead crane on rails erected the missile. I was told that recently some supposedly slow workers had been accused of sabotage. They were hanged from the crane, and their bodies were lifted and left hanging for all to see. We heard other similar stories and noted again that the German staff spoke freely with us. Surely their open speech indicated the approaching end of the war, though none of us could foresee the extreme measures taken against such discussions and other allegedly defeatist actions during the weeks preceding the capitulation. At the time of our visit to the Mittelwerk, however, many of the Germans working there must have been giving serious thought to what they would do when the end came.

The second and largest group of workers comprised prisoners collected for the V-weapons project. They wore the striped uniforms and caps of inmates in a concentration camp. I was told that the prisoners lived in fenced barracks on the outskirts of Nordhausen, close to the Mittelwerk.[3] Apparently skilled people from many countries had been assembled for this camp, and they received just enough food to survive and still perform the required work. It took no acute observation to note that an undercurrent of anticipation was passing among the workers. Everybody working in there knew exactly what was going on outside the carefully guarded mountain and its associated compounds. Clandestine radios existed, knives and other weapons had been fashioned and were hidden in the caverns, and an atmosphere of hate prevailed. The third group of people we encountered was made up of SS men, the guards with absolute power over prisoners and Germans alike. This description of the Mittelwerk, based on our observations during the brief visit and the few remarks of our informant, tells only a small fraction of the story, as I know now. I will tell more of it in chapter 11.

At last we were admitted to a locked storage area, in one of the cross tunnels, I believe. Many cartons with documents were stored there. Lists and labels clearly indicated the contents, and it was easy to find the originals of the Archive 66 series. They were not bulky; the two of us could easily carry the cartons, and we retraced our steps to the main entrance of the Mittelwerk. I do not remember the distance, but it took us some time to escape the underground works because we had to step gingerly around equipment and machinery. It was obvious to the laborers that we were uniformed outsiders, and they sneered at us in a somewhat unobtrusive way. I have never before experienced such glances of hate. Here I fall back on recollections recorded in the 1980s. There I wrote that I felt the only decent thing to do would be to rip off my uniform, put on a striped suit, and join the prisoners that all of us had put into such an inhuman situation. I have often tried to check these impressions of our visit, and I am certain about an overwhelming feeling of despair. We now tried to leave the tunnel as rapidly as we could.[4]

We must have spent only two or three hours inside that hill, but they seemed endless. It was not clear whether we could get out as easily as we got in, without being examined with our documents by SS guards. Before coming to the outer gate—all had gone well so far—we heard noises and looked back to observe from afar a change of shifts. A dense stream of men in tattered striped uniforms was disgorged from the main portal. I was frightened by this sight, and again we hurried on. From then on we had no problems whatsoever, and soon we found ourselves on a quiet road in the beautiful Harz Mountains. At that point we destroyed Wernher von Braun's orders, as we had promised. We kept the passes that proclaimed our important secret mission. Günther wanted to drive to Magdeburg to visit his in-laws. I had given up Berlin and wanted to return to Kochel as fast as possible because the Allied armies were converging rapidly from east and west. I did not relish the thought that the troops might join forces to the south of us. This would have kept us from Kochel, and it threatened us with capture by the Soviet army, which we thought was closer to us. (Actually American troops were the first to reach the Mittelwerk.) Günther prevailed: we took a detour to Magdeburg, stayed a night, ate a great deal, obtained some gasoline, and loaded the car with a great variety of food. To carry plenty of food at that stage of the war seemed to me to present a new, great risk. To army guards at a roadblock we might now look like black marketers, in addition to being spies.

Fortunately, little else of note occurred. From Magdeburg we found our old route and returned as we came, driving mainly at night. Other than seeing stray soldiers and an occasional stalled train and passing bomb craters on the road—we drove slowly—we encountered nothing dangerous. I remember that the night was not very dark, although I do not recall what phase the moon was in. We were not stopped once; in fact, few people from the nearby villages and towns were visible. Once during the night, at a location that I do not remember, we stopped to eat from our provisions and—I believe—heard the distant rumble of the fronts east and west. In the east we saw faraway flashes of light, but the corridor to the south was still wide open. Kochel looked more beautiful than ever when we arrived. The documents were deposited at the wind tunnels and later sealed in boxes. Since then, the stored knowledge has found its way into the scientific literature, and copies of the original reports can be found in the archives of many countries.

I remember events that happened before and after our trip; I cannot place them chronologically. The quiet work to save as much as possible of the installation proceeded. I do not know how much Rudolf Hermann knew of these efforts, but at any rate he did not interfere, and he generally put up an undaunted appearance.

I recall the reactions of colleagues earlier in the winter to the German counteroffensive in the Ardennes, the so-called Battle of the Bulge. A strong German force had made considerable progress in fierce fighting against the American army. In fact, at one point the American-held town of Bastogne in Belgium was encircled by the German army. The primary reason for this success late in the war was the terrible weather, which had kept American airplanes on the ground. After the weather improved, the Germans suffered under powerful air attacks and had to retreat. I thought at the time that the counteroffensive could only shorten the war by expending effective troops that would later be badly needed for the defense of Germany itself. It was, of course, well known to all at Kochel from daily observation that the Allies controlled the air. Nonetheless, even some of the scientists fell for the over-blown daily reports and believed that a turning point of the war was at hand. Others shared my view, and the offensive certainly led to the most open discussions of the war to date. Again, none of us spoke on such topics as this in the presence of Hermann and a few other people.

In late spring it became increasingly obvious that American troops would be the first to arrive at Kochel, since the Soviets had not moved beyond Czechoslovakia and part of Austria. Our hopes rose. The last two months or so prior to the cease-fire early in May—leaving out the adventure of the trip to the Harz Mountains—showed further deceptive normalcy. The wind tunnels kept running, and until the fall of Munich, mail arrived and was dispatched via our post-office box in that city. Mail was still picked up by our courier, who now used his motorcycle. The bombing of Munich was relentless, and the city was damaged to a point where new air raids only churned up the rubble. Sometimes one of us went to Munich by various means, and I recall that I had trouble identifying major thoroughfares. We kept counting airplanes during the day and going to the woods when an alarm was sounded. I recall that once when my wife and I were out on a walk, we escaped an American fighter plane that was strafing people in a field. No one was hurt in that incident.

Mysteriously, at some point I was again made the head of the Werk-schutz, a job that I did not relish and that I feared might become quite difficult at the last moment. My original thoughts of being able to help matters by using this authority withered noticeably. We now received rifles, hand grenades, and a hand-held antitank weapon, the *Panzerfaust*. Am-munition and other equipment rounded out our means of defense to the last man. The recurrent reverberation of the wind-tunnel blasts on the hills and the flood of paperwork continued to the end. Memos, reports, future plans, requisitions—often containing minute details—were exchanged with our su-periors in the army.

At one point General Dornberger appeared at the wind tunnels in ci-vilian clothes in a chauffeured car. I remembered him from Peenemünde though I had not met him personally. He was the military man who had kept the faith in the practicability of the long-range, liquid-fuel missile in spite of all the failures during the many years of A4 development. For some reason I was assigned to drive with him to Munich. I remember this trip mainly because we had a flat tire. The spare was also worn and useless, and we parked helplessly by the roadside. I recall that Dornberger, who was then wearing his general's uniform with wide red stripes on the trousers, said that a German general ought to be able to get a tire fixed. He said this with tongue in cheek, since clearly we were left to our own initiative. We bent to the task and fixed the tire with the available tools and patches. (Remember that tires had inner tubes at that time!) While we worked we had an open discussion of the status of the lost war. No *Durchhalteparolen*, or last-stand rhetoric, from Dornberger. His trip to Bavaria may have been connected with preparations for quarters for von Braun and about 130 of his colleagues, whose flight south to avoid the Soviet army must have been planned at that time.

The packing and hiding of the unique part of our equipment proceeded, and this activity must have been obvious to everybody by that time. No more visits by the party boss of the WVA, no more complaints by Hermann for not using the Hitler salute. In April, Munich was taken by the American army. No serious resistance was put up by the German army.

At the end of April or the beginning of May, I was summoned in my capacity as Werkschutzleiter to the SS casern (barracks used after the war by the American army) in Bad Tölz, the county seat, located not far from Kochel. Again I drove the old Opel through the peaceful countryside. The meadows were in bloom, the cattle grazed, and the woods showed their fresh green, impressions that could not overcome my apprehensions at this final stage of the war. At the barracks a group of men assembled in a comfortable, large room something like a conference room. Possibly twenty or so people attended, a small group considering the many organizations that were represented. Some of the participants were in uniforms of various kinds (I was not), but I did not recognize even one member of the armed forces. We were called to order by the Kreisleiter, the top party official of the county.

It was soon obvious that everybody here represented some facility: a power station, a factory, a research laboratory, a police unit, or the like. There was no calling of names or signing of an attendance list. Clearly organizational structures were beginning to fall apart. The Kreisleiter spoke at some length about directives and orders he had received from above, presumably from the Gauleiter of Bavaria working with the police units under SS control. We were to resist the enemy to the last man while army and SS units were passing through the county to set up impenetrable bastions in the Alpine redoubt. Before the arrival of the enemy, we were to destroy all bridges. Even the smallest crossings over the many streams coming from the mountains had to be blown up, together with all installations of any value to power generation and distribution, transportation, and communication. I remember the Kreisleiter as an older man who must have risen through the ranks, probably starting before 1933, the year that Hitler came to power. He tried to pass on these orders in a forceful way, including the customary threats to traitors and saboteurs, yet it was plain to us that his heart was not in it. Not a single man among the attendees at this meeting spoke up, asked a question, or made a suggestion. Nobody raised his voice to praise Hitler, recall the goals of the party, or suggest fighting on.

After the meeting broke up we were invited to participate in a guided tour of the new casern, an odd undertaking, considering the urgency of our situation. The buildings were well adapted to the Bavarian aesthetic: simple stone and wood construction with sloping roofs. At one point, we looked down into a gymnasium from a balcony. A noisy game of handball was under way. But strangely, the participants, who moved about quickly, appeared to be sitting on the floor. It then became clear that all were amputees, men in their late teens or early twenties. We knew that during the last year or two of the war the SS had begun taking young draftees and thus could still assemble strong fighting units. In marked contrast to older SS men, these young men had no political background. They found themselves caught in these units and exposed to serious accusations after the war.

On my way back to Kochel, I invited the director of the hydroelectric power plant to join me, since he had no car. We began cautiously discussing the meeting. On the one hand, we each wished to probe the other's attitude toward the imminent collapse; on the other hand, care was required because we did not know each other. It was dangerous to be openly negative before the moment when American tanks rolled into the villages. I later heard of the mayor of a neighboring village who was hanged because he had pulled out the white sheets of surrender a minute too early in order to save the place from destruction. Fortunately, it was apparent that the director of the power plant was a reasonable man, an engineer who had so far kept the plant running and who would do everything possible to keep his cherished turbines and generators from being damaged. He said that his small staff could be relied on. He wanted to stop the generation of electricity himself by throwing a few special lockable switches and by removing a few small but essential parts. I told him a little bit about the WVA and the fact that we also planned to preserve our machinery. Although we used only 2 percent of his power output, he had at least some idea of what was going on at our "water-research" station. Power was still being supplied to local villages and the spur line to Kochel, a terminus of the fully electrified Bavarian railroad system. We returned to Kochel and waited for further developments.

By word of mouth we now had accurate information about the advance of the American troops, who were moving south from Munich. More and more German units fled, passing on the main road under our windows at the Hotel Stöger. As in other parts of Germany, at the last minute a so-called *Volkssturm* was formed and equipped with outdated weapons. This last line of "people's defense" consisted of teenagers, old men, previously discharged soldiers who had been wounded, and—most dangerously for those around me—men in industry, science, teaching, and the like who had so far been exempt from the draft. Some at the WVA did not escape the fate of being drafted into the Volkssturm. I believe the top civilians were not included in this group, possibly because they were the oldest men at the WVA. Those who were drafted assembled at Kochel: they were told to bring clothing and to report to party headquarters at Bad Tölz. By various means of transportation, the unhappy little band left Kochel. As far as I know, not one of them arrived at Bad Tölz to be equipped with weapons to defend the fatherland. In the days after the armistice, which was close at hand, they all trickled back and joined us again. None of our group of about eighteen soldiers was ever called on to join a fighting unit. I suspect that the local military authorities were simply—and fortunately for us—not aware of our existence.

The next days saw a marked increase in German troop movements, all directed toward the steep Kesselberg road leading to the Walchensee. Trucks, armored vehicles, and other types of military conveyances passed by, many loaded with personal belongings. Increasingly, the hasty retreat of the disorganized units of the German army looked like a rout. The wave of traffic subsided after a few days; no bombing attack on these columns occurred near Kochel. That fact is somewhat puzzling. On one of my bicycle forays earlier in the year, I had seen a small number of airplanes bomb the railroad station at Bichl, where the Kochel spur joined the main line. A few airplanes swooped briefly below the cloud cover and neatly destroyed the station, tracks, and signals. We heard of literally hundreds of such attacks on transport, and I have never ceased to wonder why this was not done much earlier. During these last days of the war the tracks could not be fixed, and I was told later that V2 production in the Mittelwerk ceased because parts did not arrive and that the completed missiles could not be sent to the firing line either.

In addition to the retreating troops, who did not stay, civilians appeared and stayed, and they crowded Kochel even further. Some bigwigs had villas in the town, and their families showed up. Among them was Baldur von Schirach, the former Hitler Youth leader and current Gauleiter of Austria, whose wife appeared on the scene. Other officials passed through Kochel in transit to the mythical Alpine retreat, and a few Kochel residents who had reason to flee left in a hurry. For a while, great confusion existed in town.

During one of the two or three days prior to the appearance of the Americans, a long freight train tightly packed with inmates from the concentration camp at Dachau near Munich arrived at the station. The inmates represented a mixture of German Jews, foreigners from many countries, and convicted criminals. As one walked the length of the train it appeared that those who looked out the open doors were starved but not near death. There was a continuous buzzing sound of quiet excitement. The prisoners hoped for freedom, but they also knew that the SS guards still had unlimited power. The top WVA party man, with whom I had had my run-in, showed up and found out that the guards had orders to follow the insane plan of marching the Dachau inmates to the upper lake by way of the extremely steep Kesselbergstrasse. I told him that he ought to stop this scheme, which could only lead to the destruction of Kochel by the American troops, whose arrival was imminent. Realizing that his fate would then be sealed, he suddenly became quite meek and indeed seemed to accede. I do not know whether he made telephone calls or found the head of the guards. The guards were of course also much worried about their own fate; they disappeared, and the train backed out of the station in the direction of Munich.

Next we started a systematic, round-the-clock watch at the wind tunnels, because many prisoners from local jails and former slave laborers, including household help, were roaming around. No food was available for anybody, and the newly freed people began to search the farms for hidden food supplies. In short, the only remaining problem was how to get ourselves, our families, and the wind-tunnel facilities through the last hours of the war.

On or about the first of May, again with consistently beautiful weather, American troops stood right outside Kochel, after occupying the neighboring village. This date preceded by one week the armistice signed at Rheims on May 7. An eerie calm set in; Kochel looked deserted until an SS company moved in to take up positions at the edge of town near our hotel. I walked around until I found the young company commander sitting up forward in a ditch beside the road on which the Americans would come. I wanted to talk him out of defending Kochel without saying as much. After we had exchanged some general remarks on the military situation, he asked who I was, an obviously healthy young man in civilian clothes. I made some vague remarks about secret installations and pointed out that five thousand to six thousand people, mostly women and children, lived in Kochel in quarters that normally held about fifteen hundred. For whatever reasons—probably to save themselves—a little later the company disappeared.

We heard distant explosions, and one particularly loud one occurred near the foot of the road up to the Walchensee. I assumed that some of the arrangements made to blow up bridges had been carried out, leading to the detonations. (Lehnert later told me that the bridge at the foot of the mountain was indeed blown up. The blast shook Lehnert and his family, who were quartered close to the bridge.) The occupants of the Hotel Stöger brought some of their things to the cellar that ran the length of the building. I carried some items to the WVA. There we collected the weapons that we had been given to defend the installation. We piled them up in a secluded spot near the wind-tunnel buildings and covered them carefully. There was no danger of internal trouble; nobody even remotely thought of taking up arms or blowing up the equipment.

I returned to the hotel. Suddenly the strange silence was broken by artillery fire. Salvos of four rounds each were fired simultaneously at intervals of a few minutes; they were obviously light shells fired from tanks. Apparently no particular target had been chosen, and little damage occurred to buildings, as I found out afterward. One young woman who worked at the wind tunnels was hit in her upstairs room, and the young daughter of the town's pharmacist was killed directly in front of their store. I do not remember how long the town was shelled; it could not have been more than half an hour.

White bed sheets—the signs of surrender—came out rapidly; everybody had been prepared. A clanking noise was heard that my war-trained ears immediately identified as the noise made by the chains of armored vehicles moving on a paved road. Indeed, tanks and trucks appeared, and I ran out on the road. An American officer—it later turned out that he was a lieutenant colonel—walked leisurely ahead of a tank, and I went up to him. In my halting school English, I tried to identify myself, pointing out that Kochel was the home of the supersonic wind tunnels of the German army. I tried to get across that an important laboratory of great future interest was right here, and would he please come with me so that I could prove my point. I do not know what he understood, since I did not see him again; however, someone was ordered to follow me to the wind tunnels. Probably additional information about the wind tunnels reached the troops from colleagues of mine; at any rate, the next day guards were stationed there, and the place was safe. A feeling of relief came over most of us. On May 7 the armistice became a fact, and a new chapter in the story of the Aerodynamics Institute began.

N I N E

What Next?
The Summer of 1945

URING THE NEXT FEW DAYS WE LEARNED
that the war in Europe had ended with unconditional surrender by the German government and armed forces. I recall a
wonderful feeling of relief at having survived in one piece. The
American units marched toward the mountains, with only a
small group remaining in Kochel. To provide quarters, those of us who lived
in the Hotel Stöger were asked to leave and to camp in the wind-tunnel
office buildings. We stored our belongings in the extensive basement of the
hotel to make room upstairs for the military. A few soldiers were also
stationed at the wind tunnels, and we turned over the small pile of weapons
with which we had been supposed to defend the place. The instruments
were removed from their hiding place, unpacked, and set up in the wind-
tunnel testing laboratory. The hydroelectric power plant, preserved by its
director and his loyal crew, was started up, and we had unlimited electricity
because the transmission lines to the railroad and other towns had been
destroyed. Indeed, from now on we always had power, even after much of
the repair of transmission lines was done. Life in the offices was not bad at
all, and we found sufficient food. The small army detachment guarding the
wind tunnels included Lieutenant Robert Meyer, a young man who slowly
developed a serious interest in the wind tunnels. He tried his best to help us
cope with our new situation, assisting with food supplies, and later he ignored the antifraternization order issued by the American high command.

All of us were told not to leave Kochel and to stay in touch with Meyer.
Now the special situation of the members of the German armed forces came
up, a situation that I had just about forgotten. Obviously we had to be
transformed into civilians. It was Meyer's job to take us to one of the many

prisoner-of-war camps in Bavaria. It appears, however, that some higher authority was beginning to understand the value of the novel equipment and technical personnel discovered in a small Bavarian town. And so Lieutenant Meyer was charged not only with keeping an eye on us in Kochel but with making sure that the military contingent of the technical staff did not disappear in a POW camp. He loaded all of us—eighteen men, I believe—onto a truck and started to make the rounds of the camps to find a commander willing to accept us in a body while leaving us in the custody of our personal guard. Only in this fashion could Meyer be sure to follow his orders and return us all to the wind tunnels. Rumors had reached us that a huge POW camp was housed in the old Olympic stadium in Garmisch-Partenkirchen, the site of earlier winter games. German prisoners of war were supposedly jammed into this space, diseases were rampant, and it was said that deaths had occurred. At any rate, Meyer tried smaller temporary camps closer to Kochel.

During the first day of our travels, no commander was willing to accept us under the unusual conditions requested by Lieutenant Meyer. On the second day we were lucky. At a crowded POW camp in an open field, the commander was willing to accept us as prisoners and grant the stipulated immediate discharge. Under the watchful eye of Lieutenant Meyer, who did not want to lose a single man, we passed a line of German officers sitting at tables in the sunny field. A physician recorded whether we had war-related wounds or diseases, a finance officer asked whether our wages were paid up, someone else inquired about possible payments to a dependent, and so forth. It did not dawn on me at the time that this procedure was important for those German prisoners who had been wounded or who had financial claims. There are still veterans or their widows who receive pensions from the German government.

After a stay of three hours or so, we received a formal document. The tattered piece of paper in front of me at this writing states in German and English that I was discharged from an American POW camp on June 25, 1945, and had neither distinguishing marks nor any disability. The certificate was signed by Dr. Späth, a German staff physician, and by the Allied discharging officer, Gerald W. Van Eck, "2nd Lt. INF., Actg. Asst. Adj. Gen." I am still grateful to Lieutenants Van Eck and Meyer for agreeing to this exotic procedure. After about six years and six months spent in the German antiaircraft service, with two campaigns behind me, a much-shortened course of study completed during leaves, and what was most likely a lifesaving assignment to Peenemünde, I found myself once again a free man in good health. I was happy that I had beaten the odds by what I believed to have been sheer luck. With the discharge certificate in hand I received a German identification card from the Kochel town office, which was now headed by a former Communist as acting mayor.

Back at Kochel we were set to finish the assembly of our equipment and make the wind tunnels run once more.[1] From now on an unending, rather disorganized stream of visiting officers, scientists, and engineers from the United States, France, and Britain descended on us. Lehnert claims to have counted 150 such people. Most of the technical people wore some kind of uniform, and many smuggled a little food into our compound. I visited a good number of them later at their own laboratories or universities.

In parallel with our reception of visitors and the cleanup of the installation, many incidents occurred that characterized the adjustment to a new life, the transition from the Hitler period to an unknown future. This adjustment depended greatly on the nature of prior associations with political organizations during the remarkably short period—twelve years—in which all of us had lived in an increasingly totalitarian state. Moreover, the personal views of people, quite independent of party membership, office, and the like were now put to a difficult test. For example, Rudolf Hermann said to me that we now had to forget our ideology ("Jetzt müssen wir unsere Weltanschauung an den Nagel hängen"). In view of my experience with him, I could not refrain from suggesting that he ought to speak for himself. At one time he was arrested, but he reappeared after a few days. I do not know the reasons for the arrest, or who arrested him. But I found out quite recently, during my visit to Erdmann in Holland, that the German wind-tunnel people whom Erdmann saw during his visits to the United States thought this had been my doing. No matter how I may have felt about Hermann, I had nothing to do with his arrest.

We were eventually permitted to return to our old quarters in the Hotel Stöger. It turned out that our basement storage area had been thoroughly rifled, and I lost a camera. As an aside, I must say that all of us wondered how soldiers from a prosperous country such as the United States could possibly be interested in such objects as wristwatches or cameras. I could not get excited about any of the minor or major problems I encountered, however. I kept comparing my actual fate with what could have happened and with what had happened to others all around me. There were, of course, many uncertainties that arose in the aftermath of the war concerning my future work, which I still hoped would be in geophysics, possibly even in the Arctic. Where should we go after we were through demonstrating the wind tunnels? Where would we live? Could we get to Berlin in the middle of the Soviet-occupied part of Germany? The food rationing system that had functioned to the end of the war—no matter how little food was available— had broken down. Germany now experienced the hardships it had imposed on the countries it had occupied. Figuring out how to feed ourselves in the interim period of chaotic conditions did not bother me, and I wish I could have kept up this nonchalant attitude forever. At any rate, it proved useful when we moved to the United States.

I do not remember when rudimentary mail service was reestablished; however, in May and June I heard nothing about anybody outside Kochel. There was one fortunate exception: I found my brother. Because he was nine years older than I, he was not drafted until late in the war. I knew that he had last served in Prague as a second lieutenant on the staff of the army occupying Czechoslovakia, a particularly bad place to find oneself, within the immediate reach of the rapidly advancing Soviet army. An additional difficulty arose because the German army in Prague was commanded by Field Marshal Ferdinand Schörner, a wild and brutal officer who had risen through the ranks through political connections. My brother later told me that Schörner was known to shoot soldiers who walked in the wrong direction.

My brother's wife and small son had been evacuated in 1943 from Berlin to Unterammergau, a small Bavarian village adjacent to the more famous Oberammergau. We had occasionally visited them by train from Kochel. I now enlisted the aid of Lieutenant Meyer, who, of course, had a jeep to drive me there. Within sight of the snow-covered Alps, we drove through the lovely countryside. Fields, woods, and villages were all untouched by the war. Finally, as I got out of the car on the green—the cow pasture of Unterammergau—I saw my brother walking toward me on his way to the store!

We embraced. It seemed to me a miracle to find him so quickly, in fact to find him at all. It turned out that, in contrast with his comrades, who had fled from the Soviets in a westerly direction in hopes of becoming POWs of the Americans, he had walked alone to the east, toward the approaching mass of the Soviet army, whose advance troops had already reached Prague. He had ripped off his insignia and carefully hidden his Soldbuch in his boot. As he passed the advancing Soviet columns with their small horse-drawn wagons, friendly soldiers occasionally waved to him. After walking some distance, he turned north to reach the border between Czechoslovakia and Germany. On this portion of his flight, he was largely alone, detouring the Czech villages. Back inside German territory, farmers fed him and permitted him to sleep in their barns. Next he turned west to the Neisse River, the demarcation line between the American and Soviet troops that had been fixed at previous Allied conferences. From the eastern bank of the river he shouted to American guards posted on the other side until he found a friendly soldier who told him to come over. My brother swam across the narrow river and surrendered. The understanding soldier took him prisoner and delivered him to an American POW camp, where he was released within a few weeks. Most of the German soldiers who had fled from Czechoslovakia to the west were overtaken by the Russians, and those that were captured in the area by Americans were later turned over to the Soviets. This procedure had been previously arranged, and many of the prisoners spent years in captivity. Reunited with my brother in Unterammergau, I spent some time with him and his family before returning to Kochel. After the trains began running—the minor connecting lines between villages had barely been touched by air raids—we visited more regularly until my brother, a lawyer, left for a job in the Ruhr area.

Soon after the arrival of the American troops we heard that the train from Dachau that had left Kochel in the direction of Munich went only twenty to thirty kilometers. The remaining guards fled, and the prisoners were free to leave their freight cars. They quickly spread in all directions. Because Dachau was one of the concentration camps that was also used as a prison for criminals, great excitement broke out among the population. At the same time, many of the forced laborers from foreign countries, who had worked in factories, in stores, and often in private households with children, began to roam the Bavarian countryside. Those who worked in households, mostly young women from the Ukraine, were often treated like members of the family. They stayed on, and other forced laborers from Russia did the same in view of the lack of transportation and the unclear future for them in the Soviet Union. A feeling of uncertainty took hold among the population, a strange feeling that was mixed with relief at the end of air raids. The surrender forced party officials to leave their posts as mayors, county officials, policemen, and so forth, stopping all governmental activity on the local level for a time. All this added to the general confusion.

The events that preoccupied and overwhelmed me were of a different nature. Soon the German radio functioned alongside the continuing German-language broadcasts of the Allies. As I mentioned before, I was quite aware of the persecution of Jews, Communists, and politicians of all parties in the Weimar Republic who had stood up to Hitler. But only gradually did I realize the enormity of the destruction of the large Jewish population of Germany, citizens who had lived there for many generations. It became known that the same fate had befallen Jews in the countries under German occupation. One new name of a concentration camp after another appeared, names I had never heard of. Later, photographs of the murders in the camps became available. A few of these camps were in Germany, but most were in the east, primarily in Poland, and my geographical knowledge could not even place them. The slow awakening to the mass murders of Jews and others extended over the whole year. Since I describe a good part of my life in these pages, I believe that I ought to answer the question of what I did or did not know about what is now called the Holocaust.

As a student and later a soldier on leave in Berlin—a large city with many channels of information—I was exposed through my father's friends to many stories about the government's doings beyond those officially known. Before being drafted I traveled several times to foreign countries and heard much that was not known in Germany. Early in 1933, I had heard for the first time the term concentration camp (*Konzentrationslager*) in conjunction with Oranienburg, a small town close to Berlin. It was there that politicians, people in the arts, and others thought to oppose the new regime were incarcerated. Many of those detained were set free after a while, but nobody was willing to speak about the experience. We knew well a man who had been there for several months but said nothing about his treatment. I further knew that a similar installation existed in Dachau. Rumors also circulated that terrible things were occurring in a large, well-known children's hospital in the Mark Brandenburg, not far from Berlin. There physicians were said to kill children who had birth defects or other infirmities. I have spoken already about my later foray into the ghetto in Minsk, where I learned that German Jews had been deported to this place and subsequently were taken away by truck and never returned. In addition, as a soldier one could occasionally observe the aftermath or hear of the actions of the special SS units that ran the administration behind the German lines. Such actions included the execution of Russian partisans. I once saw a number of men who had been hanged from a single tree, all carrying signs announcing that they had been caught while taking part in actions against the occupiers. This account covers all my recollections of the existence of concentration camps and related matters.

Obviously, as a Berliner who grew up surrounded by Jewish friends—I had been one of the few non-Jewish children in a private Jewish elementary school—I was aware that Jews were slowly disappearing from the city. This was apparent in spite of my extended absences from Berlin at boarding school until 1936. Again, it was a slow process, but it was particularly apparent with respect to friends of the family in the theater and in the movie industry. On the other hand, we also knew of many who had emigrated, including a number of Jewish students at the boarding school.

We did not know about the extreme difficulty of finding countries willing to accept Jews when it was still legally possible and relatively easy to leave Germany. When the Jews were forced to wear a yellow star sewn to their clothes, two facts could be observed by those who kept their eyes open. For one, it became apparent that many Jews were quite poor, contradicting by their appearance Hitler's theory of the rich Jews. Clearly, poor people could not possibly afford emigration. The second fact was that their number kept decreasing. I will omit the many experiences of my father, who could occasionally help owing to his position in the theater and the movie industry, and I will not speak about my own minor efforts along these lines. I heard after the war that a surprisingly large number of unregistered Jews had remained underground in Berlin throughout the war. This group depended on the clandestine collection of food from other Berliners, among them my family.

Again, it is essential to remember that the persecution and murder of Jews took years. The so-called Final Solution, the decision to eliminate the Jews from Germany and the occupied countries, was made as late as 1942, as we now know. At that stage—remember the Stalingrad disaster in January 1943 and the much-increased air raids—the war preoccupied everyone. This situation aided in maintaining the secrecy of the mass murders. I believe that during the war only a small number of Germans had ever heard about the Final Solution, Auschwitz, and the gas chambers.

This darkest chapter in the history of Germany became known to me bit by bit during 1945. Much of it I learned later by reading the historical literature and visiting the sites of the concentration camps, the remains of the Gestapo headquarters in Berlin, and similar places. It may be difficult for someone who did not live at the time in Germany to believe all this, but it is the truth of my own experience.

I do not remember the date when communication with other places in Germany became possible, but finally we heard from our relatives. My wife's family lived in Essen, which was in the British zone of occupation. We found out that the second of her three brothers had died at the very end of the war, most likely even after the armistice. The submarine on which he served as engine officer was sunk by British airplanes in bad weather while on the surface. The boat was on its way to surrender according to the instructions issued to submarines by the Allies. Possibly confusion caused by the weather led to a misunderstanding. All others in her family, including the brother who had been too young to be drafted, were in good shape. I in turn received news from the other side of Germany, the Soviet zone of occupation. My father had stayed in Berlin up to the last day of the battle for the capital of Germany. During the final fighting in his area of the city, his house was jammed with women from the neighborhood. He had made a film—I believe in 1929—in the Soviet Union, and now he attached a poster or photographs with Russian script—I'm not sure which—to the door of the house. He stationed himself nearby and succeeded in keeping the rampaging Soviet soldiers from entering the place.

Shortly after the armistice was signed in Berlin, the Soviet Marshal Georgy Konstantinovich Zhukov, the conqueror of eastern Germany, ordered General N. E. Bersarin, the commander of the city, to speak with my father. The history of the interview with Bersarin, as I now know, was as follows. Shortly after the fall of Berlin, a group of German Communists who had emigrated to Moscow after Hitler came to power returned to the city. This group later provided the nucleus of the government of the German Democratic Republic. They heard that the Soviets intended to set up an organization of those involved in cultural affairs to reorganize the theater, the movies, art, music, and later the newspapers. The German group proposed my father as the head of this undertaking. Although he had always stayed away from political identification, even in the period of the Weimar Republic, he had always opposed the Nazi party, particularly the party officials in the arts. Moreover, he was well known in Berlin, and so General Bersarin followed the German group's advice and offered him the presidency of the Kammer der Kunstschaffenden (Chamber of Those Working in the Arts), a job that he felt he had to accept. After the arrival of the three Western Allies in Berlin, the respective heads of their sectors, which formed an island in the middle of the Soviet zone, approved the establishment of this chamber and the appointment of my father. It is hard to believe in retrospect that already in May, the month of the capitulation, plays and music were being performed under the most primitive circumstances, and soon plays forbidden under Hitler were being performed to sellout audiences.

On the practical side, my father's position guaranteed that his house had glass put in the windows, that the roof was fixed, and that some food was provided. On the other hand, he had to make difficult choices in appointing the people to get all this cultural activity going. It was hard for the occupying powers to understand that a party member was not necessarily a Nazi or that the converse could be true. Later the Allies had civilians assigned to help in the work. Some had previously worked in the field and had left in time to escape Hitler's persecution. They knew my father and helped much, but still he was overtaxed, and his health suffered. Nevertheless, a dream of his, sustained and often discussed during the Hitler period, came to pass. He had hoped to survive the war to be able to play Nathan der Weise (Nathan the Wise), the sagacious old Jew in Lessing's famous play of the same name. Indeed, he played this role sixty times in Berlin and elsewhere prior to his death in 1948. In the meanwhile, my mother—my parents were divorced—whose apartment in Berlin had been destroyed by bombs, had moved to

Prague, the city of her birth, where a younger brother and other family members could help her get settled.

The life of showing our stuff in Kochel became routine. Rumors popped up that the institute might be moved to the United States and that some of the staff members might be taken along. Apparently discussions went on behind the scenes among the British, French, and Americans, but nothing definite could be found out by us. It seemed, however, that the institute was safely in American hands. We now had more time to turn to other pursuits. When the trains began to run we visited friends in Munich. I recall that in the rubble left from the air raids I could not detect the location of one of the main streets of the city. All I found was a narrow footpath zigzagging up and down the piles of debris. I had lunch in a famous restaurant where a thin soup was served in the basement of the ruins of a once-grand mansion.

Many other events made life interesting. I was again arrested. A higher officer of the Counterintelligence Corps, if I remember correctly, asked me in a friendly way to join him in his jeep. We drove to a place that I cannot remember, where I was courteously interrogated. Apparently the Gauleiter of a state in eastern Germany, most likely Pomerania, was named Wegener, a common name in these parts. It was obvious that the skinny twenty-seven-year-old recently released POW could not possibly be the person they were looking for, and after a short time I was released. Other episodes were related to the fact that I was in obvious good health and relatively tall. American soldiers were still on the lookout for members of the SS. Those SS men who had served in a fighting unit had the symbol of their blood group tattooed in their armpits, and this indelible identification was used to discover SS soldiers. After I took off my shirt, the American soldiers looked for the tattoo and released me.

In spite of the antifraternization orders issued by General Eisenhower after the discovery of the conditions in the concentration camps, we found that our relations with Lieutenant Meyer remained unchanged. One night he and a captain friend of his knocked on the door of our room at the Hotel Stöger. I can still hear him whispering, "Peter, food, food." The two had shot a deer and carried the beast to our door. Our facilities were hopelessly inadequate to accommodate this plentiful supply of meat, and we had to live on without game. But Lieutenant Meyer gave me a truly important gift, his copy of the fifth edition (1943) of *Webster's Collegiate Dictionary*. I can still see him carefully scratching out his name from the inside cover with his pocketknife. From then on I never used a German-English dictionary, and thus I avoided the pitfall of translating from the German while writing, or even when speaking. My high school English provided a sufficient base for this method of learning another language, and I am still grateful to Lieutenant Meyer for his conversational and lexical assistance.

At some time during my stay at Kochel, I had been told that Baldur von Schirach had a villa in our resort town. As I mentioned above, he was the original Hitler Youth leader and later the party head and regent of Austria.[2] It was also said that a painting by Brueghel graced a wall of his living room. Many high officials of the government liked to cover their walls with art taken from museums in the occupied countries. Since von Schirach's villa had been taken over by the Americans, I mentioned this fact to Lieutenant Meyer and suggested that he have a look and make sure that the painting would be preserved. He did not say much then, but one day he came to tell me that an official from the Netherlands had arrived to reclaim the picture for his country, from which it had been stolen. Apparently those who pointed out stolen paintings and other works of art were usually rewarded for their detective work, but Lieutenant Meyer, who had missed receiving an award, was sad that I had not told him that the painting might be worth a million dollars or so. To end my relation of such little stories, I note that people apparently did search for hidden gold near the planned party headquarters on the shore of the Walchensee, yet nothing ever was found. Had the gold been dumped in the water, the depth of this glacial lake surely would have made retrieval impossible.

One day Werner Heisenberg arrived by bicycle to join his family at Urfeld. Heisenberg, who had received the Nobel Prize in physics in 1932, and his team of nuclear physicists obviously knew that in principle a bomb could be built to release the enormous energy of nuclear fission. So much discussion arose in the years after the war about the extent and objectives of the German work, and in particular the intentions of Heisenberg, who had been close to some of his counterparts in the West. We now know that his group did not succeed in producing a chain reaction or in figuring out how an atom bomb could be built. At any rate, the arrival of one of the best-known German physicists added to the scientific interest of the area.

Rumors continued to circulate about the fate of the wind tunnels, and consequently about our own future. It became clear that the United States had a priority claim on the Aerodynamics Institute, because Kochel was in the American occupation zone of Germany. American soldiers were in place, and most of the visitors studying our work were Americans. I am not sure of the exact sequence of events, but the administrative director of the WVA, Herbert Graf, suddenly disappeared with some employees in the direction of France. It was said that certain papers went along with them. Nobody I knew joined him, and none of the scientific staff of the WVA followed him. Then, to our great surprise, Rudolf Hermann disappeared without any previous announcement. We heard that he was on his way to the aeronautical laboratories in Ohio at the Wright Field air base, which became one of the major research centers of the new U.S. Air Force. I still wonder what made him decide to leave friends who dated from the founding days of the institute in the late 1930s and to dissociate himself from the Aerodynamics Institute, the major achievement of his life. I did not think then or later that he believed he had to avoid difficulties based on his political history. Although I had had my problems with him—certainly nothing compared with Erdmann's experience—I could not believe that he had serious reasons to worry about his past. In the subsequent years in the United States, I had no contact with him. It was strange that I did not encounter his name in the technical literature, but he may well have worked at the government laboratory on problems that had a security classification. I wrote to him once to clarify a point related to early technical problems of the Peenemünde wind tunnels, but he did not answer my letter. Since I understood that I had been something of a thorn in his side, I did not pursue the matter. After Hermann left, Kurzweg became the informal but obvious head of our activities, a situation that I gladly accepted.

Several possibilities for our future arose. For one, we could start work in Kochel on aerodynamics research directed by American scientists. But a renewed operation would make sense only if urgent technical problems remained that had a direct bearing on the war against Japan, since it was thought that the war would continue for some time. Another possibility was that, after some haggling among the services, one of the branches of the U.S. War Department would be presented with the Aerodynamics Institute. The wind tunnels and their instrumentation might be shipped across the Atlantic in carefully numbered cases—the way medieval castles were moved to the new world—to be attached to an existing laboratory or to become part of a new one. If this course were chosen it would make sense to take some of the German staff to the United States for a certain time to speed up reassembly. Such a move might not, in fact, be a huge operation, for the buildings, the large vacuum tank, and other components of the institute, such as electronics and machine shops, could be left behind.

By the middle of the summer many of us were becoming restless about the uncertain future. It appeared that a single person had to take over, become familiar with the Aerodynamics Institute's history, operations, and capabilities, determine the facility's usefulness for the United States, and make concrete proposals to the defense establishment about the future of the institute. Detailed technical knowledge about previously unknown types of wind tunnels could not be imparted during short visits to large numbers of curious people from many different places. Fortunately, such an individual appeared in Kochel in the person of Fritz Zwicky, an energetic professor who wore three hats. As I now know, he was a technical representative of Team 183 of the Combined Intelligence Objectives Subcommittee (CIOS), a large group of advisers formed in August 1944 by the Combined Chiefs of Staff. He was also the director of research of the Aerojet Engineering Corporation in Pasadena, California, and, most important, a professor of astrophysics at the California Institute of Technology. Zwicky spent most of May, June, and July at Kochel, and during this period he put together an eighty-nine-page report on the Kochel activities and their background that strongly urged moving the wind tunnels to the United States.[3] He also put his proposal in the context of American research and development of supersonic missiles. He noted that on his arrival at the wind tunnels he had counted 180 scientists, engineers, mechanics, and administrative personnel, of which half were found to be unessential. According to Zwicky, the military boss of Team 183, a Colonel O'Mara, had permitted those that were not needed to "return to their point of origin." O'Mara came from the southern United States, and even those of us who recalled some high school English could not understand him at all. Zwicky further noted that the equipment was in good order and that the original reports on vellum were found. These were, of course, the reports that Günther Hermann and I had retrieved from the Mittelwerk.

Although I did not see his report at the time, I saw a great deal of Zwicky. It took me a while to sort out the difference between him and casual visitors and to grasp that his job at the wind tunnels went much beyond curiosity. It was clear from the outset that he was different. To start, he was a powerful man originally from the German-speaking part of Switzerland who spoke Swiss German with us in a booming voice. Nothing escaped his attention during his inspections of the equipment and during the extensive interviews he conducted. He slowly developed a strong belief in the quality of the technical find made by the Allies at the foot of the Bavarian mountains. I got to know Zwicky well during several extensive interviews and many conversations with him aside from the business at hand. Whenever he disappeared—to report elsewhere, I suppose—I began to worry about what might happen at Kochel should he not reappear. Once when he returned from a trip I heard him say something to the effect that Hitler and Stalin had upset the world and that he had to set it right again. This task may have been too much even for Zwicky. But such remarks, of which there were many, did not diminish his broad scientific and engineering knowledge and his singular effort to get the wind tunnels to the United States. In fact, he was the first person from whom I heard concrete proposals about this idea, including the need for some of us to go along. It then dawned on me that he might be the one person whose proposals would be heeded. His somewhat volatile temper occasionally made things a little complicated, but everything he did served the same goal. He even involved some of us in details of his plans, though I do not know how seriously he took our thoughts.

In spite of the reassurance offered by Zwicky, I became restless at Kochel. The stream of visitors thinned and rumors still circulated, but no clear indications about the future were forthcoming. Burdened by these uncertainties, I decided that the time had come to take action rather than waiting around. First, I wanted to see family and friends beyond Munich. Next, I had to be prepared to accept or reject an offer to move to the United States with the institute should such an offer arise. I had often felt the urge to leave Germany, quite aside from the sadness I felt about my country during the last twelve years. At boarding school I had lived with many students from other countries, to whose stories I listened. I had traveled outside Germany prior to the war, and, after all, my mother and her family were from Czechoslovakia. As a soldier I had seen France and Russia; both in their different ways had much impressed me. And on my Spitsbergen trips I got to know Sweden and Norway. On the other hand, I wanted to pursue geophysics, primarily in the Arctic—a path that I had been inspired to follow by my relatives Alfred and Kurt Wegener and by my mentor Albert Defant—so I needed to learn what my job opportunities in Germany might be. In the summer of 1945 such matters were obviously uncertain, to put it mildly. No real government existed, and legal and other arrangements were very different in the different zones of occupation. The larger cities were destroyed, the universities were closed. My prior contacts in geophysics were probably dispersed. Little transportation was available, and no telephone service functioned, but still I felt that I had to act.

At first I thought about seeing my father in Berlin and my mother in Czechoslovakia. On reflection, I decided that surely I could not see my mother, because travel to a foreign country, even by hitchhiking or bicycle, could not possibly be brought off. But how about Berlin? I had received letters from my father with hints that a trip to see him ought to be postponed. It slowly dawned on all of us that the Soviet zone of occupation, which had to be traversed to get to Berlin, was something quite different from the areas occupied by the Western Allies. My older sister had crossed the Soviet zone, and after returning to her apartment in Würzburg in the west, she wrote about searches by guards, demands for papers, and the like. In addition, we heard from sources that appeared reliable that the Soviets had lists of names of Peenemünde people who would be interned once they were found. Indeed, the Soviets assembled a group of scientists and others from Peenemünde who had remained in the east, then took them to the

Soviet Union where they spent many years working on its rocket program. This group was augmented by scientists in many other fields, in particular atomic and nuclear research, and surely this fate was the last I wished for. Therefore, my wife and I made less adventurous plans to start with a visit to her family in Essen.

Since I was still tied to Kochel by an understanding with the American military, my wife took off alone on a hitchhiking venture beyond Munich. Before the war she had owned a small automobile, a pretty Ford Eifel convertible for two people and some luggage. At the beginning of the war private cars, such as that of my father, had to be turned over to the government for use by the army. My wife had avoided doing this with her convertible, and nobody ever seemed to look for it, possibly because it was useless for military purposes. At any rate, the car was hidden in a barn near her family's vacation home outside Darmstadt, south of Frankfurt. She made it from Kochel to the barn and found that after five years the car was still in the same place. With some help, she got it started and drove to Essen.

Remaining in Kochel, I wrote a number of letters, and to my delight the mail worked sufficiently well for me to receive relatively rapid answers. I had correctly assumed that Defant, my professor, had left Berlin before the end of the war to go to his home in Austria. Professor Möller, who had helped me so much during the preparation of my thesis, had remained in Berlin. She responded with sadness, telling me about the terrible conditions in Berlin and the destruction of the university. Since I had explained my situation, she strongly urged me to go to the United States if at all possible. To resume work in geophysics appeared impossible to her, and Defant, who had become the rector of the University of Innsbruck, would have Austrians as his assistants. She thought that Germans could not even enter Austria at this stage. I also received a response from the head of our Spitsbergen expeditions, Herbert Rieche, who sounded much more optimistic. He, as well as Ernst-Günther Triloff, a member of the 1937 and 1938 expeditions, had been drafted by the weather service of the armed forces. Unlike me, they both had long finished their studies at the time of our Arctic forays. Rieche, who had lost his young wife and a baby in an air raid, had been sent by the army to establish a weather station on the major eastern island of the Spitsbergen archipelago. This part of Spitsbergen was not exposed to the Gulf Stream, so the climate was much rougher than that which we had encountered in the western island. The station operated year-round through the dark Arctic winter, and supplies were delivered by submarine. Triloff, whom I considered to be by far the most gifted member of our expeditions, was similarly sent to work at a year-round weather station on northeastern Greenland. The final supply ship, which was supposed to deliver materials with which to build a hut, had been sunk by the British navy, and the men had to spend the winter in an ice cave that they dug in a glacier. (When I visited Triloff in recent years he repeated how much he had enjoyed his second Arctic venture, including the accommodations.)

Rieche had already contacted Triloff to propose further research in the Arctic, and he wanted to include me. He had found some interest on the part of the British weather service, I believe, where the German efforts to gather atmospheric data during the war had been evaluated. It appeared to him that a new expedition into the north might be supported by the British. I thought that this initially attractive plan was much too vague and could not possibly succeed so close to the end of the war. At the same time, although I did not wish to forsake the slight hope of staying with the wind tunnels, the uncertainty about employment was troublesome. My savings had dwindled, but I knew that I would always be able to get a job as a forest worker, a trade that I had learned during my Arbeitsdienst and that I had enjoyed even at Kochel. In the meantime, nothing of interest turned up, Berlin was out of bounds, and the best solution seemed to be to stay in Kochel and hope for the best.

In part because of the rapid ups and downs of the decision-making about the wind tunnels and our future, the order of events is not quite clear in my recollection. Zwicky and O'Mara disappeared, and the local army commander took us over rather loosely. Soon after Zwicky left, or perhaps even overlapping his tenure at Kochel, Mott-Smith, an American scientist in uniform assigned to the navy, arrived. Mott-Smith, who had worked in our field of gasdynamics as a theorist, was one of the first people to try to calculate the structure of shock waves. Such waves are extremely narrow disturbances that cause sonic booms, among other effects. Normally one calculated the thermodynamic state of the air, such as its temperature and pressure, on both sides of the practically discontinuous shock wave. But it was new to study the internal structure of this discontinuity to determine how the gas changed its conditions so drastically in the narrow transition zone of the discontinuity, which is much smaller than the thickness of a hair. Here we finally had a person authorized to state specific plans saying that the wind tunnels had been assigned to the U.S. Navy and would be transported to the United States, and that selected German personnel would be asked to go along. Mott-Smith was charged with supervising preparations for the transfer. It was interesting to find out that the navy, with its carriers and submarines as launching platforms, was most interested in aircraft and missiles, including the practical application of research in supersonic flows. I discovered only after the war that the air force was not a separate, coequal branch of the services but rather the air division of the army. In view of the importance of aircraft during World War II, this seemed to me a strange arrangement. A separate air force was soon founded, however, though many special types of aircraft and missiles were still attached to the other services. The new air force also wished to acquire people from our group who were familiar with supersonic aerodynamics, and Hermann, who had already left, was of course the first.

In the meantime, the war in the Far East had come to an end. The first of two atom bombs was dropped on Hiroshima on August 6; the Japanese surrendered on August 14, with a formal treaty signed in Tokyo Bay on September 2. These developments put an end to one of the possibilities for us: the resumption of research at Kochel. Under the new circumstances that would make no sense at all. I now thought that if I was really selected to go along, a short absence from Kochel—if permitted—could not make much difference. Indeed, I readily received leave, if you can call it that, and was equipped with an essential piece of paper stating that the U.S. government was interested in my existence in order to serve it as a scientist. Therefore, after my wife returned with the little car, we took a trip of about three weeks to see friends in Heidelberg and proceed to the Ruhr area in the British zone. I had little trouble, with my American paper, getting gas and oil at the filling stations run by the army on the autobahn. Oil was important, because the car's shadow existence had not improved the smoothness of the interior of the cylinders. Supplies to run the car became scarce in the British zone, and I remember that after showing my paper to a British sergeant I may have been close to another arrest as an impostor. As in Bavaria, the discrepancy between the bombed cities and the untouched countryside and small villages and towns was stunning. The timing of the events that followed our trip can be described more accurately, based on letters that I wrote at the time and on diary entries.

After we were settled again at the Hotel Stöger, this time in more comfort augmented by a rug, books, and other items brought back from Essen, there was little left to do. The packing of the wind-tunnel equipment required little attention, and soon the area around the railroad station was littered with wood cases and big pieces of equipment, such as pumps and electric motors. The Kochel vacuum tank was, of course, left behind, as well as items that could be readily obtained overseas: electric transformers, switching gear, and the like. Now we were told that about twenty wind-tunnel people were being invited to come along, and I was supposed to be among them. We were asked to fill out the exceedingly cumbersome so-called *Entnazifizierung* (denazification) forms written in German that practically all Germans in the American zone were presented with. I did this as well as I could, though many questions either were not applicable to me or were obscure. In addition to personal data, detailed information was required on all organizations of which one had been a member prior to 1933 and during Hitler's time. Based on these forms and subsequent hearings, people were arrested, were kept from performing their jobs, were not permitted to enter certain fields of activity, or were let go. The premise of the forms was, of course, that everybody was a Nazi, and only by this complicated procedure could the degree of involvement of the individual and the subsequent punishment be established. In retrospect it is obvious that—aside from the most blatant cases—no proper matching of the degree of guilt and the actual behavior of a person could be determined. I heard no response after I handed in my forms, no interview took place, and to the best of my knowledge no others were asked about me. The same was true—as far as I know—of my colleagues. It seemed to be clear, comparing our situation with that of others outside the wind tunnels, that special treatment was being afforded us.

With only twenty people on the list—substantially fewer than the fifty that Zwicky had proposed—I assume that many of those who had stuck it out at Kochel were disappointed. It turned out that only about nine of the twenty actually went with the tunnels; most of the others received assignments to other laboratories in the American armed services. Although in October everything appeared to be set, no personal contracts with details of the operation were available. According to Lehnert, the French made use of the uncertainties at Kochel to snatch away a few more people. He also claims that a wind-tunnel employee who was ready to leave on a truck with his entire family and household goods was stopped at the last minute by an American guard. Since I did not know any of these people and was gone part of the time, this story bypassed me. Those who were selected for other places than the navy laboratory were being moved temporarily to a place near Munich called Camp Overcast. This was the first time that I heard the code name Overcast. My friend Eckert was in this group, and I was sorry to be parted from him.

There is no question that during the last phase of Zwicky's work all of us had complete freedom to leave Kochel to go anywhere we wished. Moreover, even the initial stay at Kochel had been virtually voluntary. It had soon become clear to the Americans that we had made serious, even risky efforts shortly before the end of the war to save the wind tunnels because we felt a responsibility for them. Many of us had hoped for a future with them. A few had left; for example, one gifted aerodynamicist had inherited a farm. A few like me gave serious thought to a return to previous work. In this complicated situation we were therefore only "asked" to remain in Kochel and not put under some form of arrest. Even after being chosen for the overseas move, however, I still felt uncertain in every way. It was said that our papers—including the political form, I assume—had been sent to the navy in Washington. The officials there obviously had the last word, and we had no idea what they might decide about any of us. Some complained about the uncertainty, but of course we were wholly dependent on the pace of the bureaucratic process.

Finally, toward the end of the year, contracts were submitted to those who were to rebuild the tunnels. I found myself in this group, and after careful reading of the proposed arrangements and discussions with those who presented the contracts, I decided to follow the supersonic wind tunnels to the United States. The contracts were written for a half year, allowing an extension if needed. We were to work in the area of Washington, D.C., where the wind tunnels would be rebuilt at a naval research laboratory. Our families had to stay behind, but, depending on how the first year worked out, they might be able to follow. Since nobody could leave a family in Germany to fend for itself, families were to be housed at Camp Overcast. This "camp" turned out to be the houses of the noncommissioned officers attached to the extensive barracks of the dissolved German army in Landshut, a pleasant small town about fifty miles northeast of Munich. The town was untouched by the war, and it seems that the inhabitants of the houses had been moved elsewhere. Although we had experienced numerous advantages in the past, I thought that this was the first time that many had been truly inconvenienced by us. Fortunately, this was the only occasion of the sort. Our families would stay behind and receive our salaries in German marks, a sum that would be DM 11,000 per year. In the United States we were to receive a per diem of six dollars and free lodging. A certain number of pounds of luggage was allotted, to be shipped separately. We would also have to agree to remain in a designated area, since we obviously had no normal immigrant status. The navy would, however, "take care of us." I quizzed one of the officers to find out what the fuzzily designated "area" might be, and he answered that it might be the city of Washington and its surroundings.

At no point in our negotiations was pressure exerted to accept the navy's proposal and contract. Clearly we were wanted, but we were free to decide our own future. The attitude of the United States about our move was the opposite of that of the Soviet Union. The Soviets simply took those Peenemünde people and others to their country for long periods of time. There followed later—until today, in fact—discussions in the United States about the pros and cons of the whole move, as we shall see.

We now packed, gave up our Kochel quarters, and got ready to move. I recall a particularly wonderful Christmas in Kochel, the first Christmas after the war, with snow and midnight mass, which I regarded as a ceremony of thanks for having survived the Third Reich and the war. On January 1, 1946, we arrived in Landshut and were shown to quarters that were vastly superior to those at Peenemünde and Kochel. The contract had included heat and food for the families, many with small children, so that we could leave without worry. It was great to wander around Landshut, which was considered one of the few jewels of the Gothic style in Germany. The narrow, pencil-like steeple of one church dating from the early fifteenth century was the tallest steeple in Bavaria and, at 436 feet, the tallest brick steeple anywhere. The center of town looked much as it did in the sixteenth century, and the few days I spent there were like a vacation in an exotic place.

On January 11, the nine (or so) wind-tunnel people, including Kurzweg, Lehnert, and others mentioned previously, were ready to start. In the morning we took a last walk to the town's medieval castle to look across the Bavarian landscape, which is quite flat in that part of the state. At 4:30 P.M. we boarded a bus that brought us to the railroad station, and from there we took a train to Munich. There we spent the night at a place called Hotel America, and the next day another train took us toward the west on the first leg of the great voyage to America. In many ways we were just another group of German immigrants who had traveled over the years in the same direction, starting in the middle of the seventeenth century. But I believe that, for better or for worse, no Germans had ever traveled before under such strange circumstances.

Moving to the United States: Project Paperclip

UNKNOWN TO ME AT THE TIME, I WAS NOW A part of Project Paperclip, the newest code name for the program to bring German scientists to the United States. Previous code names like Overcast had become generally known, for example from the designation of the Landshut quarters as Camp Overcast. It is said that the name Paperclip originated when an official at the Pentagon affixed a clip to the papers of those Germans who were to be invited. When the move occurred, the military agencies were eager to keep it under wraps. Soon after our arrival, however, the controversial project was discussed in many forums; it continued to be a hot topic for the first few years of its existence, and remarks about the project still appear occasionally.

On January 12, 1946, we left Munich on a perfectly good express train. We moved slowly, however, probably in view of technical problems remaining with the railroad system. After passing Stuttgart we crossed the Rhine River, which was once again Germany's border with France. Via Strasbourg and Nancy, we reached Paris early on January 13. We arrived at the Gare de l'Est and had to transfer to the Gare St. Lazare for further rail travel to the Atlantic coast. The transfer between stations was done on an old-fashioned bus with an open platform at the back from which I immensely enjoyed the sights of Paris.

This was the second time I had seen Paris. I had spent one day there shortly after the armistice in France at the end of June 1940. At that time we were stationed at an airfield near Paris, on the lookout for British aircraft that never appeared. German troops other than those assigned to the relatively small detachment of the Paris command were not permitted to enter the city. But a few of us received fake papers stating that we had to testify at a court-martial in Paris. This ruse, invented by our commander, gave us a glorious day in the city. We drove there in a truck. At the entry to the city we were checked to make sure that we were not carrying any weapons. We looked at the major sights, ate a wonderful lunch, and ended our visit at a nightclub. I still remember running through deserted boulevards late at night to our meeting place, barely catching our truck to return to the air-field. Driving across Paris in a bus four and a half years after my first visit, seeing the beautiful city that had experienced little damage—with auto-mobiles driving around and well-dressed people in the streets—it seemed as if there had been no war.

We boarded another train and left Paris at 2 P.M., arriving in Le Havre after a seven-hour ride. A truck picked us up, and we drove to the old citadel near the inner harbor. There were huge rooms in the massive walls of the fortification, and about thirty participants in Project Overcast—including our small wind-tunnel contingent—were put up in one of these cavernous spaces. Single beds had been set up, the place was well lit, and I looked forward to a pleasant waiting period until a ship was available to take us to New York, as we were told to expect. We ate in a mess, and I was over-whelmed by the quantity of the food. I do not know how I would judge it today, but after the food of the past two years or so, the cuisine struck me as luxurious. I tried to be careful not to eat too much and to stay away from the heavy stuff. Some of our group were less careful in adjusting to the new diet and suffered dire consequences. Every morning and afternoon during our wait we could leave our grotto to take walks on the walls of the fortifications and enjoy the view over the city and parts of the harbor.

I read a great deal, as had been my habit even when I was on active duty in Russia, and in a letter to my mother I remarked how well I felt. Some of the inner group around Wernher von Braun—who had preceded his team to the United States—traveled with us. Ernst Stuhlinger, a physicist in a leading position at Peenemünde who much later became von Braun's biographer, gave a talk on January 18 describing the basic idea underlying the atom bomb and what he called atomic machines that might be built. This talk and many discussions about the future helped to while away the waiting period. On the following day a Lieutenant Hobbins addressed us on the subject of the United States, but I cannot recall whether he discussed history or how we ought to behave ourselves. Clearly, we were under careful supervision, and strolling freely through Le Havre was out of the question.

The great day of our departure arrived. On January 20 at 4 P.M., a truck brought us to the harbor, where we boarded the *Central Falls Victory,* one of the many relatively small ships that with other types of vessels still sailed between the continents, ferrying troops home to the United States. We were assigned a large cabin for about forty men. There were bunk beds, and I quickly spied an upper berth directly below a louver that I took to be a fresh-air vent. When confronted by mass accommodations, food lines, and the like, I could fall back on the experience gathered during my years in the service to secure as much comfort as possible in a given situation. Two young officers—who pointedly did not fraternize with us—were assigned to guard us; I believe they lived with us in the large cabin. We were permitted to leave the cabin only for meals, which we ate in the officers' mess. (I particularly remember plentiful fresh apples.) We were told not to speak with anybody else on the ship. Finally, at 10:50 A.M. on January 21, the ship left the pier to start what turned out to be the worst sea voyage that I have ever experienced.

We were barely outside the harbor when the ship began to forget that it was supposed to move on a great arc toward New York; rather, it moved up and down, performed corkscrewlike movements, and veered from side to side. I remember that nearly all of us rapidly became seasick. In addition, the vessel creaked loudly when it rose above the waves and fell back down on the water. Images of broken steel ships began to haunt me, and I was scared. The Atlantic in January is a tempestuous sea, as we found out. After a few days my seasickness disappeared, but many of my colleagues remained in bad shape. We were told that many of the soldiers in the large hold below us did not recover during the trip. Although I had been correct in assuming that my louver brought in fresh air, I needed to get out. So I simply escaped to the deck and found a place where I could hold on and observe the tall waves and the ship's motion. The weather did not let up. In fact, we heard that during one twenty-four hour period the ship made no headway at all toward the west but simply staggered at a right angle to the oncoming waves. Finally, the weather cleared, and very early on February 3—after an Atlantic crossing of fourteen days—we saw the lights of Long Island shining toward us, and were now permitted to go on deck to watch our arrival.

It was a cloudy day, and slowly the tall buildings of Manhattan emerged. I was not particularly impressed by the sight, and I am still amazed at my lack of wonderment. Today, when I take the Staten Island ferry just to experience this view, I marvel at the dramatic appearance of the tall buildings. There have, of course, been major architectural changes since 1946, but I still ponder what was the matter with me when I saw all this for the first time. The *Central Falls Victory* docked on the Hudson River side of Manhattan and began to disgorge its passengers. We were told that the arrival of ships filled with returning soldiers was always published. Because the designations of the units and their hometowns were given, journalists from all over waited on the pier. The reporters had obviously not been informed about the additional strange passengers aboard, whose arrival was carefully kept secret. We had to remain on board until everyone else had left the ship.

Finally, much later and in the dark, we went down the gangway and made our first contact with the soil of the United States. Our luggage appeared, and we boarded a chartered Greyhound bus. I recall that I was fascinated by traveling on an elevated highway and viewing the lights of New York, awed by the size of the place. Much later, we crossed a wide river by ferry, and the city of Washington appeared. Although it was quite late, many lights were on there as well, and we passed over a bridge into Virginia. It was now past midnight, and we stopped on an open road in a dark forest. Military buses showed up, and we transferred to these vehicles with our luggage. Next, we entered what was obviously a military camp, where rooms were assigned to us. By now I was very tired, in large part because of all the new impressions, and I quickly fell asleep.

After getting up the next morning, I noticed German-speaking young men who cleaned up, made beds, and the like. It turned out that prisoners of war worked in the camps, and I hustled to make my own bed. I thought that it could just as well have been me serving a group of German scientists, and surely I could not view my fellow soldiers as servants. This was yet another reminder of how lucky I had been. On February 5 we underwent a physical examination by several doctors; I felt fine and was declared healthy.

Next, we were interrogated based on questionnaires we had filled out in Germany and also, I believe, forms we filled out in the United States. This information was possibly enriched by separate investigations that—unknown to us—had been made of our lives. I remember clearly that I was questioned primarily about three points. When asked to state my religion in the forms, I had put in none. I was asked whether I had been baptized, and indeed as a small baby I had been baptized by a Protestant minister, the husband of an aunt of mine. He had spread his baptismal paraphernalia on my mother's white grand piano. As a child, I was never taken to a church for a service. However, I visited many churches out of an interest in history and architecture, interest that has stayed with me. Lessons in religious history were offered at my boarding school, but those of us who were Protestants could in addition take religious instruction from a minister in the nearby town of Holzminden to prepare for our confirmation. As a teenager, I was puzzled by the basic tenets of Christianity. I quizzed the minister on the actual meaning of confirmation, and on what one had to believe to become a true Christian. After a period of indecision I concluded that I simply did not share the required beliefs and that I could not take the major step of confirmation. The interrogating officer, to whom I did not divulge any of these details, suggested that all I had to do was to write "Protestant" on the form like everybody else. This simple solution to what I considered a serious question did not appeal to me. It reminded me of the attitude of some of my Protestant classmates, who thought I was crazy not to look forward to the first suit with long trousers, the watch, and other presents customarily given to newly confirmed boys.

Another point concerned my political past, such as it was. In the fall of 1936, after my service in the Arbeitsdienst, I had to join the Nazi student organization (Nationalsozialistischer deutscher Studentenbund), a requirement for matriculation at the university, just as it had been for the Arbeitsdienst. This fact appeared not to be taken seriously by the officer who spoke with me, who surely worked under carefully thought-out guidelines that weighed different political involvements. But what about my Hitler Youth membership at the boarding school? I spoke about the events at the school that had led all of us to join and the director's idea that our joining was essential to saving the school. The interrogating officer suggested I state that I was forced by the school to join. I pointed out that I could have quit the school, gone home to Berlin, and transferred to a public high school, where membership in the Hitler Youth was not yet required. Again I did not want to budge. Obviously the team questioning all of us, many at great length in fact, had been instructed to play down our political pasts as much as possible. My involvement had been unimportant in this context, but I simply wanted to tell it as it was. I wondered about the many Paperclip people who had been in the party and related organizations. Where would a line be drawn?

Finally—and I believe only in the forms given to us in Virginia—the famous question arose concerning whether one had ever been a member of an organization whose aim was the overthrow of the U.S. government. I had, of course, put down yes, to the consternation of my interviewer. I pointed out that the German air force, of which I had been a member, was an organization whose obvious aim was the destruction of the United States. Since this was not taken seriously, I was released and could spend the remaining days in the camp in Virginia as I chose. The time there was not unpleasant; twice we even went to movies on the base.

On February 11 we moved to the Anacostia Naval Station, at the southern edge of the District of Columbia. We were quartered in a Quonset hut. When I wanted to take a walk, I was told to stay in the hut. I made myself unpopular by saying that such a restriction was not stipulated in our contracts, but I did not get anywhere. Two naval men were now responsible for our well-being; both were helpful and had—by sheer coincidence—German names whose spelling I am not sure of. Strangely, neither Kurzweg nor Eber showed up at the new camp.

On the following day we made our first trip in a van through the city of Washington to the Naval Ordnance Laboratory (NOL) at White Oak, Maryland, close to the small town of Silver Spring, just outside the District of Columbia. The navy had acquired a large area to build a new laboratory covering many disciplines. The laboratory had previously occupied cramped quarters at the Naval Gun Factory in Washington, and now a great expansion was planned. Our workplace was in a large new warehouse where temporary cubicles that were open on one side had been furnished with desks and office equipment. At the other end of the warehouse, architects and draftsmen of the construction company had their drafting boards and files. Thus we were thrown into a practically normal work environment without any restrictions in our dealings with secretaries, technical people, and the crew that was building the laboratory. On February 14, Kurzweg and Eber finally appeared, and on the same day we were even taken to a concert, where I particularly enjoyed a Mendelssohn symphony, a piece of music that could not be performed in Germany during Hitler's time.

The plans for the wind tunnels went ahead, with the building that would house the equipment patterned after the one at Peenemünde. Again, the center was taken up by a huge spherical vacuum vessel; this time, however, it was welded rather than riveted. Our group's work involved assisting in the design of the buildings and suggesting improvements to the wind-tunnel equipment. The generators, transformers, switches, and all other electrical installations, as well as pipes between the German pumps and the tunnels, were designed by the contractors. In the meantime, I had plenty of opportunities to think about the future. It soon became clear to me that I did not wish to be involved in the setting up or operation of the wind tunnels. Rather, I wanted to work on hypersonic flows, the field of the future. Erdmann had laid the groundwork of one aspect of testing at extreme speeds with his first wind-tunnel experiments at Mach numbers much higher than those attainable with previously existing equipment. Here the problem of the condensation of the working medium—the air itself—had arisen; it was a new problem that I wanted to tackle (chapter 6, note 7). Moreover, the understanding of hypersonic flows encompassed extreme flight speeds in the earth's atmosphere, an environment very different from that of standard wind tunnels, and such problems as the entry of meteorites into the atmosphere arose. To work on any of these problems I would need to increase the meager knowledge of chemical physics that I had obtained during my shortened studies. I began this remedial education by consulting various books, and even when I mentioned my plans I was left to my own devices, in part because we really had no organizational structure for some time. I could take full advantage of the freedom left to us.

Malcolm Rice, the representative of the contractor who was overseeing the construction of the laboratory in White Oak, often worked in our building. I was much impressed by him and enjoyed listening to his many stories. He had held the same job during the construction of the National Gallery in Washington. The facade of that building is made of Tennessee marble of a pinkish hue. Before they were transported to Washington, the marble slabs had been laid out on flat ground. Rice had a platform built from which he looked down on the future wall. He had the slabs moved around until the colors were matched to provide a uniform shading of pink from the ground level to the roof. Next the pieces of marble were numbered and shipped to Washington to be installed in the desired order. An architect who worked for the construction company, Dick Collins, became my closest friend during the seven years that I spent at NOL. He also taught at Catholic University in Washington. After I was free to move around, we saw much of each other, and I was accepted by his family and his friends.

Scientists from the laboratory were permitted to take us to their homes after signing for us at the gate. Newspapers were available, and we could occasionally go shopping with one of our guardians. Parties were arranged, we were taken on sightseeing tours as far as Mount Vernon, where we admired George Washington's house, and all in all the navy was most liberal, under the circumstances. In fact, one of the historians writing on Project Paperclip notes that the navy was indeed the most relaxed service by far with respect to the treatment of Paperclip scientists. For one, we had the substantial advantage of being only nine Germans, in contrast to the relatively large groups in Texas with Wernher von Braun and at Wright Field in Ohio. Later, when the wind tunnels were back in operation, the Paperclip group made up only a small fraction of the total NOL staff. Moreover, we had a large city at our door, in contrast with the desert where von Braun's group worked. These facts permitted a much better accommodation to American life and afforded a better chance to learn the language, the system of government, the civil service, and similar things. For example, I found going to the movies very helpful in learning English. We were driven by marine guards in a station wagon to either the Silver or the Seco movie theater in Silver Spring, and in addition to the films, the marines vastly enriched our vocabulary. In March, our old friend Bob Meyer visited us to see how we had fared and whether the equipment packed under his guidance had arrived in good shape. I was glad to see him again.

At the end of May we were able to leave our tight navy quarters to move to the grounds of the White Oak laboratory. The long, unpleasant ride through rush hour traffic was over, though we had recently switched from the smelly van, whose exhaust fumes found a way inside, to private cars driven by scientists from NOL. The large area acquired by the navy included several buildings of various sizes and purposes that existed prior to the purchase. Gerd Eber and I were assigned to a small farmhouse that had been renovated. We each had our own bedroom, and there were also a workroom and a kitchen. For the first time since our departure from Camp Overcast in Landshut, we enjoyed privacy in our living arrangements. Life became much more pleasant; we could cook a little and save by taking care of our own laundry. I remember that I never achieved Eber's ability to iron a shirt in less than ten minutes. Spring had arrived, and the fenced, partially wild laboratory grounds invited exploration on long walks. The only obstacle to greater enjoyment of the partial freedom was the financial restriction imposed on us by the per diem of six dollars. Other than the most urgent purchases of clothing, every penny had to go toward buying food and sending it to relatives and friends in Germany. Our American coworkers helped, for example, by collecting bacon fat in coffee cans that we taped shut and mailed.

Further good news was received in July. Our families would be permitted to join us in January 1947, a year after our arrival. This news was followed in October by a second pleasant surprise: we could roam freely in Washington and the neighboring part of Maryland. (On the other hand, no public transportation was available at NOL until 1947, although we did not have to be signed out any more.) No matter how relaxed the regulations had been, this change made us breathe easier. The granted escape from our confinement implied that the navy took us to be reliable participants in the work of NOL. It was obviously now assumed that we would not disappear from the scene. During this period I began to consider seriously whether to stay permanently in the United States. This consideration was not connected with avoiding the tough circumstances I would encounter in Germany were I to return but was based on the increasing attraction that I felt for the new country.

Indeed, in January 1947 our families joined us. We were driven to the wonderful Union Station in Washington to meet the train from New York. Private quarters had been assigned at a naval base south of Washington. This location introduced us to a new round of commuting. At roughly the same time, we started to receive a civil service salary that I assume was reasonably commensurate with our age and professional experience. (Later I found out that my view was not shared by many Paperclip scientists elsewhere.) It was now essential to be able to get around, and with the help of friends I purchased at very low cost a 1940 Ford coupe, a black two-seater of wholly indeterminate mileage. This car ran quite well for two years or so. Next, we were encouraged to find private living quarters on the open market. After an extensive search in unknown territory, we settled on a one-bedroom apartment in Washington, at 5504 Seventh Street at the corner of Kennedy Street in the northwestern part of town. The rent of forty-five dollars—low even for the standards of the time—probably was influenced by the view from the living room of a brick wall ten feet away. Moreover, the apartment was on the second and top floor, under a flat roof that was not insulated. I never really got used to the climate of the nation's capital and wondered why the founding fathers did not choose a cooler spot. However, I was happy to be free, to live in a neighborhood that was quite pleasant at the time, and to face the challenges of assembling some furniture and kitchen equipment from flea markets and the like and settling down to a normal life. In addition, I bought some hand tools and built tables and shelves myself, another pursuit that has largely stayed with me. After all, I knew that things could only get better. In little more than one year, after a move between continents and under the most unusual circumstances, we were immersed in the normal life of the United States, a country with rules and forms of behavior very different from those we had known.

From this point on, the chronology of events again becomes a little jumbled in my recollections and contemporary notes and letters. On the personal side, a wonderful first vacation in Virginia and West Virginia took place later in 1947. In 1948 my presence in the country was formally legalized. With the navy acting as sponsor, immigration papers had been prepared. I traveled to Niagara Falls by train with a young naval officer in November 1948. There I walked across the Rainbow Bridge to meet a U.S. consular official on the Canadian side. After filling out additional papers and undergoing more formalities for three hours or so, I walked back to the U.S. side, showed my immigration visa to the border officials, and went through the gate. At that stage I became a truly free man again, a situation that made me feel good. Having permanent immigration visas, documents that our families had received upon their entry, opened the way to international travel. In 1951 I flew for the first time across the Atlantic, in a sleek Lockheed Constellation. After a twelve-hour flight from New York, we landed at Prestwick, Scotland. During a layover I talked to a shepherd whose sheep kept the grass short, who spoke of the hundreds of American aircraft that had left from Prestwick to bomb Germany during the war. This was the first of many trips to Germany, Europe, and other parts of the world.

At some time in 1949 we rented a little house in the countryside outside the District, and finally—again with the help of friends—we purchased a small house on the Briggs-Chaney road in Montgomery County, Maryland, a short drive from the laboratory. Ours was a quiet road with few houses, a road that originally connected the farms of Mr. Briggs and Mr. Chaney. It is now a fully built up area with a huge tree behind the house we lived in, a tree that I planted more than forty-five years ago.

At White Oak, construction proceeded rapidly, but I do not remember when the tunnels started to run again, because I was wholly occupied with plans for a small hypersonic wind tunnel that I hoped to be able to build. At any rate, the period during which wind-tunnel work was interrupted was much longer than that associated with the move from Peenemünde to Kochel. An organizational structure took form as well. Raymond J. Seeger took over as head of aerodynamics, the wind tunnels themselves, applied mathematics, and related areas. Seeger, a physicist whose Ph.D. was from Yale, had played an important part in mobilizing science for the navy during the war. He attracted major physicists to work in such areas of fluid mechanics as the behavior of shock waves. Some of these individuals had approached Seeger to ask what they might be able to do to help the war effort prior to the time when many of them went to Los Alamos to work on the bomb project. In addition to organizing groups mixing the American and German scientists at NOL, Seeger used his range of acquaintances as a reservoir of consultants and seminar speakers. I learned much from these contacts, and the field of science really opened up for me.

Under Seeger, the Aeroballistics Research Department was formed. This terrible name stems from the navy's tradition of artillery. Kurzweg led what I believe was called a division, and I was responsible for Hyperballistics Subdivision III. It appeared that I might indeed be able to start work based on my plans for hypersonic research. In addition, Seeger wanted our group to build a shock tube, a then-rare device with which the behavior of shock waves could be studied. We built this apparatus with the indispensable assistance of George Lundquist, a young electronics engineer, because the events to be observed took place in microseconds. At the same time, it still took a lot of talking and persuading to get permission to build the small hypersonic wind tunnel that I had proposed. It appeared essential to me to build such a research facility to demonstrate that very high Mach numbers could be reached in a nozzle. I had allies in Kurzweg and others, and finally we succeeded. A group of engineers and technicians needed to be put together, and it included two Germans, the gifted mechanical engineer Edmund Stollenwerk, without whom we could not have done the job, and Eva Winkler, who did the intricate work on the thermodynamics of the wind tunnel. The other members—including George Lundquist—were Americans. We were helped considerably by Jeanne Beitzel, who was good at performing the endless calculations that were required. The slide rule was still king, and electromechanical calculators worked laboriously and noisily.

Our relations with the scientific, technical, and clerical personnel of the entire laboratory evolved gradually. We made use of the auxiliary services of mechanical and electronics shops, offices that ordered supplies and equipment from commercial sources, and the like. Relatively soon, and in varying degrees, our English ceased to be a handicap. But I will never forget the moment when an American suggested that we call one another by our first names. One day in the corridor Zaka Slawsky asked me, "What is your handle?" and I did not understand his meaning. Zaka was a physicist whose specialty was quantum mechanics, and although he was Jewish, he ignored my background as a German who had grown up under Hitler, served in the Wehrmacht, worked at Peenemünde, and so forth. Several Jewish refugees from Germany, who had fortunately gotten out in time, worked as applied mathematicians in Seeger's department. They avoided us. Nothing seemed to me more natural at the time and even today in view of what Germans had done to the Jews of their own country and much of Europe. I strongly believe the idea of collective guilt to be a misguided notion, yet the shadow of the Holocaust is a burden that my entire generation has to accept. (Here I move into a subject to be taken up in the last chapter.)

There were, of course, many other exceptions besides Zaka to the blanket condemnation of us. One member of my group, a theoretical physicist, was working for his Ph.D. on a new approach to a tough problem related to the condensation of air. His dissertation supervisor—a professor at a local university—was one of the major German physicists who had found shelter in the United States just in time. The three of us talked about the research, and the professor helped me just as he might any other person struggling for a solution.

My first publication was an abstract in the *Physical Review,* a journal of the American Physical Society. The abstract related to a talk I gave at a meeting of the society, which had been introduced by Seeger. Shortly after that event he proposed some of us for membership in the society, and we were accepted. This was a daring step for Seeger, because the whole Paperclip program was being severely attacked by other scientific societies.

Scientific travel slowly began. The entire German group was asked to visit the Ames Laboratory of what later became NASA. We traveled to California on an old DC-3 piloted by a reserve officer. I remember a pleasant overnight stop in the middle of the country at an air base where the drinks in the bar cost next to nothing. The Ames Laboratory was fascinating, and we also saw nearby Stanford University and San Francisco. A similar DC-3 trip that included the worst thunderstorm I have ever encountered in the air took us to Florida. But I also accepted invitations on my own to many places, such as the General Electric laboratories in Schenectady, New York, where rocket development was taking place, and the Universities of Minnesota, Virginia, and Toronto. I entered a consulting agreement with Ohio State University in Columbus. This required air travel on many Saturdays to Columbus, where I tried to assist with the planning of an aeronautics department. I remember that the discussion at lunch always centered on football, a game that was completely new to me and has remained more or less a riddle to this day.

To complete the list of academic events that occurred parallel to work at NOL, I note that our hypersonic-tunnel work was presented at meetings of professional societies. The fact that we had reached a Mach number of 8.3 in a wind tunnel free of disturbances was indeed noteworthy. In addition, I taught courses at NOL under the auspices of the University of Maryland. Later, Francis Clauser, possibly my most important adviser at the time, who had founded a department of aeronautics at the Johns Hopkins University after the war, asked me to teach a course on gasdynamics that met on Saturdays. I enjoyed the drive through the beautiful countryside between Washington and Baltimore, an area that now seems to be one continuous town, taking on a dangerous similarity to Los Angeles. During my NOL period, I received a number of job offers from universities. Teaching had been my ultimate goal since my student days, but I heeded Clauser's wise suggestion that I do research for a while, an activity that can be most easily performed at a good laboratory like NOL. If I went to a university, Clauser said, I should join it as a full professor and avoid the rat race and uncertainties of promotion. I planned to follow his advice, although it was a course of action that had been much easier for him to take in view of his brilliant career as an aerodynamicist at the Douglas Aircraft Corporation during the war.

In 1953, I accepted an invitation from Frank Goddard to join the Jet Propulsion Laboratory (JPL) operated by the California Institute of Technology. This major university-operated laboratory, which at that time was supported by the army, combined the prospect of freedom of research with the availability of wonderful facilities. It was proposed that I lead an aerodynamics research group and a team planning a large, continuous hypersonic wind tunnel. In particular, the flexible nozzle system pioneered by JPL for supersonic wind tunnels would lend itself beautifully to such a new facility. My decision to leave NOL was strengthened by the fact that advancement on the GS levels of the civil service appeared to be related to administrative position, and I did not wish to become a division chief at White Oak, the next step in line. And so I accepted the offer to start work at JPL in the spring of 1953, after seven years with the navy. Ralph D. Bennett, the technical director of NOL, asked me to stay, pointing out that the navy had recently arranged promotions for people who did not take major administrative jobs. I thought that seven years was enough, however, so I set out for the West. Bennett was an outstanding scientist and laboratory director who left us unusual freedom; he had led the move to White Oak and the expansion of activities. He retired as a research director of the Martin Marietta Corporation in 1966, and he died in 1994 at the age of ninety-three, as I noted in his obituary in the *New York Times*. In sum, I have only good things to say about the navy's handling of research, a statement that includes support from the Office of Naval Research that I received much later at Yale. Unfortunately, in view of the many fine publications that emanated from the laboratory under the name of NOL, it was subsequently renamed Naval Surface Warfare Center, or NSWC, adding one more letter to the acronym.

This is not supposed to be the story of my whole life. Since the last vestige of being a member of Paperclip vanished with my separation from the laboratory where the original wind-tunnel people of Peenemünde were employed and with my becoming an American citizen on August 26, 1954, in Los Angeles, the story that I wanted to tell has now been told. In the fall of 1960, I joined the faculty of Yale University as a professor, following Clauser's advice. There remain many underlying problems and questions concerning Peenemünde and the relocation of German scientists to the United States, and I shall add some remarks on this subject in the last chapter.

ELEVEN

Looking Back

L ONG BEFORE I THOUGHT OF WRITING DOWN
my recollections of the war years in Germany, I asked myself about
the meaning of this period in my life. Although I am not given
to philosophical soul-searching, I felt that the general curiosity
about this time made it important for me to record my memories of
Peenemünde. I was also encouraged to do so by my three American-born
adult sons. Moreover, the older I became, the more I thought about the past,
a trait that is surely common among those getting on in years. For example,
why did I accede to the draft in November 1938, and how did I feel about
service in an army whose goals I did not share? When no further postpone-
ment of army service became possible, I did consider leaving Germany,
possibly to join relatives in Czechoslovakia. They spoke German, but the
only Czech that I could speak consisted of asking for ice cream, and I did
not see how I could make a life in that foreign country. Obviously I did not
anticipate the invasion of Czechoslovakia, which soon brought me, as I
mentioned at the beginning of this narrative, to the potato field near Berlin.
At any rate, being fundamentally optimistic, I realized that I had to stay in
the country that was my home and hope for the best.

I hated every moment of my first half year or so as a recruit at the
antiaircraft barracks in a suburb of Berlin. I was close to tears at night when
I heard the electric bus driving past, a bus that would have taken me in
twenty minutes to my father's house. My opposition to the so-called basic
training was not based on the strenuous exercise that we were exposed to,
because I was in excellent physical shape, nor was my revulsion connected
with any political belief. I just could not take the mindless discipline of
going through various modes of marching, waking to night alarms in order

to clean the barracks, and learning the obviously useless stuff one was supposed to learn. Later, in the field, the actions required for survival depended on independent thought and initiative, none of which was taught in basic training. I suspect that similar criticisms may be directed to other armies.

Starting in the summer of 1939, life improved. With many other units from the Wehrmacht, my battery drove for maneuvers to the province of Silesia, on the eastern border. Of course, none of us knew that Hitler planned to attack Poland later in the year, and this show of force turned out to be a part of his plan. The weather was fine, and the countryside was beautiful. We slept outside or in quarters made available by the local farmers, we ate well, and the mindless discipline melted away. In contrast with life in the Berlin barracks, the whole trip seemed like a vacation, and it was the last peaceful time of my military life.

This period ended abruptly in September 1939, when the Wehrmacht attacked Poland and my unit was moved from the Baltic to Berlin. From then on, strangely enough, the relative freedom experienced during our summer maneuvers was maintained during the actual campaigns. When things were rough, the primary motivation of soldiering took over—a feeling for the people one lived and fought with—rather than any ideological impulse. Hitler's name, or even the glory of the fatherland, was never invoked. Everybody hoped to survive, and enthusiasm appeared only when one of the rare leaves became imminent.

During training courses for prospective corporals or sergeants that I was ordered to attend, a lesson called history usually showed up on the schedule. Obviously, history was taught along the party line. The only outright political speeches filled with adoration of the Führer that I can recall occurred at the reserve officers' school in Bernau near Berlin, which I attended for a few weeks in the winter of 1940–1941. One of the instructors, a first lieutenant in the flak who was a member of a well-known family of industrialists and whom I remember as handsome and well educated, was a committed follower of Hitler.

Another example of ideology concerns a run-in I had with a major who headed a flak unit of three batteries, equivalent to a battalion. I was assigned to his staff after I became a lieutenant. It was my job to contact the infantry, armored unit, or whatever group our flak was assigned to with the primary purpose of fighting Soviet tanks. I drove around on my huge BMW motorcycle, saw a great deal of what was actually going on, explained to regimental or divisional staffs what the flak could and could not do, and lived an independent life. The major in question called my friend Petri—the staff physician—and me to his quarters in the previously mentioned city of Orel. Many members of our unit had just seen *Jud Süss*, the most viciously anti-

Semitic movie ever produced in Germany. Petri and I had been disgusted with the film, and our conversation was overheard by the major. He said that he had considered reporting us to higher authorities, but because of our importance to his staff, he had not done so. He admonished us to behave. There were, of course, other party followers, mostly older people, yet all in all we lived free of political pep talks and the like. In part, I believe, the war was too grim; nobody invoked Hitler when we buried the dead whom we had known well.

In sum, serving in the Wehrmacht without hoping for the often-cited Endsieg—and after the 1941 attack on the Soviet Union, not seeing any remaining chance of it—made for mixed emotions. By the end of July or so, only about six weeks after we crossed the Bug River, my unit was stuck beyond Smolensk. Too many had been killed or wounded, and our equipment was in terrible shape, so we could not participate in any further advance toward Moscow. In December, Hitler declared war on the United States shortly after Japan, a German ally, attacked Pearl Harbor. At roughly the same time, the Russian winter set in, with disastrous effects on the German advance, because the army was not equipped to withstand the cold. The failure of the Wehrmacht to succeed in Russia before the winter stopped all movement, combined with the prospect of the United States sooner or later entering the war in Europe and the fact that major German cities were exposed to increasingly serious air raids, reinforced my bleak outlook. Many others at the front or at home became equally apprehensive.

My own future changed drastically when, after my last return to the eastern front, military orders brought me to Peenemünde late in May 1943. At that stage I assumed that fighting in the field was over for me; I had escaped in one piece, something that I had taken to be most unlikely. A strenuous period of more than three years had ended, a period when I was often scared to the point that I thought I could not take it. In parallel, however, I had often felt a strange fascination in observing the unfolding events. With these feelings behind me I entered the radically different world of scientific research in a field of which I knew little.

Let me add a few thoughts about my stay at Peenemünde. Much has been written since the last world war on research in science and engineering in totalitarian societies. Because scientists are often rather independent characters, how could they work under such circumstances? First I must confess that I did not even think about such a problem while I worked at the wind tunnels. My joy at having escaped from the eastern front, coupled with the new experience of doing research in a well-equipped laboratory, outweighed all other reflections. It was once thought that original research and development would be impossible under a government like that of Hitler, or for that matter, Stalin. I believe that by now this notion has been laid to rest. The quality of research performed during the war years in Germany and—most notably—in the Soviet Union after the war was based on scientific curiosity and the hope of discovery. Government service or ideology—with obvious exceptions—did not play a major role. At the same time, the threat of losing their exemption from the draft was a consideration of the civilian scientists and technicians, providing an additional inducement to work hard.

An interesting comment concerning the possible motivations of General Dornberger, the force behind military missile research, was made by another soldier, a British air chief marshal who thought highly of the officer who pursued rocket development from the earliest days: "Was it the urge to travel through space—now being described by a leading astronomer as 'bilge'—was it the almost religious feeling that leads scientists to carry out research which appears to have no useful purpose, or—more likely—was it the military man's desire to make a bigger and better weapon than that possessed by enemy nations? But these efforts made Britain face a weapon for which there was no reply." Rather than assuming that Dornberger was driven to his work by ideology and adherence to Hitler, de la Ferté speaks of complex, apolitical motivations.[1]

In discussing motives, I ought to add the truism that the views of a selected group of scientists (here used throughout as a generic term including engineers) do not represent the views of a cross section of the general population. An overwhelming majority of Germans undoubtedly would have voted for Hitler in a secret election—perhaps after the armistice in France in 1940, when Great Britain remained the only adversary. A remarkable number of scientists—among them Ludwig Prandtl—retained their independence to the point of taking risks in the defense of their field, its practitioners, and their professional societies.[2] Some of this independent spirit was reflected in Peenemünde, as inferred from discussions with people one knew well.

In the previous chapter I wrote about my own experience with Project Paperclip in Germany and in the early days of my residence in the United States. I mentioned that shortly after my arrival in Washington it became apparent that such a program could not exist without provoking serious controversies. I shall now report some of the events that I did not know about, ranging from the search for Paperclip candidates to the political problems in the United States. To understand more, I looked at some of the literature on Paperclip and closely related events, such as the capture of German scientists by the Soviets.[3] What happened during the last days of the war and right after the armistice?

So far I have used the code word Paperclip loosely. In fact, Lasby and McGovern point out that Overcast was the first term used for the attempt by the Joint Chiefs of Staff to bring German scientists to the United States. Lasby notes that in July 1945 the Overcast staff "envisaged the temporary exploitation of a maximum of 350 chosen, rare minds, to help in the war against Japan, and to aid postwar military research." The word *exploitation* was dropped from government documents when movement to the United States actually started. Lasby writes that "increasing Soviet intransigence" led the State-War-Navy Coordinating Committee to propose Project Paperclip as late as the spring of 1946, after I had moved to the Washington area. The plan proposed inviting one thousand "specialists" and their families to enter the United States and, implicitly, offering them citizenship. This plan was approved by President Harry Truman half a year later. The cold war was beginning to affect many matters of policy, and thus one aim of Paperclip was a "denial" of the immigrants to other nations, especially the Soviet Union.

An even earlier action, termed Crossbow, involved the frantic search for German science and engineering projects of all kinds, ranging from new submarine drives, jet engines, materials, nuclear research, and possible development of an atom bomb to the V1, aircraft laboratories, and finally the V2, including the supersonic wind tunnels. American scientific teams followed the front lines closely even before the German surrender, looking to discover new German technologies. After the last shot was fired, many sites were searched. The searches were carried out independently by the different services, so there was little coordination. The American race to capture German scientists was complicated by other nations' pursuit of the same goal. The Soviets did their part in the east, as we shall see, but in the west roamed teams from Canada, Great Britain, France, and several other countries. Even Yugoslavia entered the fray. Hundreds of searchers were deployed by the United States, the prime mover, with keen competition among the services. The second largest group, the British, were not included in a joint approach, despite carefully worked out prior agreements.

In the hunt for the V2 it was essential for the American teams to be the first to reach the Mittelwerk. At that time, the Allies knew where the V2 was being mass-produced. An American tank unit arrived on April 11 at the small town of Nordhausen, Thuringia, near the Kohnstein, from whose tunnels Günther Hermann and I, with the aid of von Braun, had retrieved the original wind-tunnel reports. The U.S. troops discovered the remnants of the concentration camp complex Dora with its dead and starving prisoners. Colonel Holger Toftoy of U.S. Army Ordnance, whose mission was to secure and exploit the missile production site, the Mittelwerk, ahead of the Soviets, and who later became the military mentor of von Braun's group in America, was informed in Paris about the eagerly awaited occupation of Nordhausen. He was prepared to retrieve as many completed V2 rockets and parts as possible, to be sent to the United States. In this endeavor he was joined by a civilian group led by Richard Porter of the General Electric Company. Under contract with the army, Porter had headed a missile project called Hermes, and he later showed up at Kochel. After I came to the United States, he invited me to the General Electric laboratories at Schenectady, my first trip away from Washington. Toftoy's special mission was delayed until the end of April: remember that all this took place prior to the armistice.

Anxiety among the Americans rose with the rapid Soviet advance, and since the agreements at Yalta had ceded the area of Nordhausen to the Soviets, it was essential to remove everything as soon as possible. The tunnels and storage areas inside and outside the Kohnstein yielded V2s in various states of completion, as well as parts for the assembly of additional units. The material was secured through the incessant work of a group of dedicated soldiers. By chance, a freight train was stuck in the area. Railway personnel were found, and the largest shipment of German equipment to the United States during the entire Crossbow and Overcast operations was set into motion. Eventually, four hundred long tons of disassembled V2 parts were moved from the Mittelwerk area via train and Liberty ships to New Orleans, and from there to Fort Bliss, Texas, at the northern edge of El Paso. Late in 1945 and early in the spring of 1946, the 127 members of von Braun's core group came together in New Mexico, and work began. Stuhlinger writes about the assembly and firing of the missiles, including high-altitude research, through which Goddard's dream of the application of rockets was finally realized in the United States. These firings in the southwestern desert culminated in the moon landing in 1969.

While the American soldiers were securing the hardware at the Mittelwerk, it was learned that the original documents concerning the A4 design, testing, and the like were hidden in another abandoned mine nearby. With ingenious detective work and the aid of the Peenemünde people still in the Nordhausen area, the documents were saved shortly before British troops moved in. Fourteen tons of papers were shipped to the American zone. In retrospect, it seems ironic that the seeds of the cold war were planted before the armistice and that the British, of all people, were excluded from the important technical findings on the first guided missile in history—the missile to whose devastating effects they had been exposed. I find that my friend Rees from Haus 30 was involved in tracking the stored papers. The poor fellow, whose intellectual independence had kept him far away from Hitler's ideology, was housed for safekeeping in a local prison, like some of the other people who had worked at Peenemünde In contrast, at Kochel, aside from the brief arrest of Hermann, all of us were actually quite free to come and go as we pleased, after initially agreeing to stay nearby for a while.

Of the Paperclip literature that I have read, Lasby's work is by far the most informative, solidly based on historical research. After years of effort, Lasby finally broke down the unreasonable security classifications of government documents relating to Paperclip imposed by the Pentagon and the State Department. In his discussion of the discovery of the wind tunnel, I found some interesting remarks by Zwicky and Millikan relating to our work. Zwicky was pleased with the "great efficiency" and the "unusual spirit of cooperation" of the wind-tunnel group. He supposedly "cracked the whip" to get us to work at eight in the morning, finish up the reports, and so forth, while I am certain that I remember a much more leisurely schedule. Another matter that escaped me at the time was Zwicky's complaint that the army guards had "greatly hindered" his efforts by breaking into the laboratories, snatching souvenirs, and the like. Zwicky attributed the cooperation of the scientists to a "lack of loyalty to any political doctrine" based on the "severe lack of education, insight, or even interest of these scientists in both internal and international politics, sociology, and economics. Almost to a man they had been completely fooled and impressed by Hitler's apparent creation of a financial and economic *perpetuum mobile*." I find these remarks truly puzzling; my frequent discussions with Zwicky included general topics, and I never noted any antipathy toward us. It appears that he intensely disliked Germans, however (see, e.g., David H. DeVorkin, *Science with a Vengeance* [New York: Springer-Verlag, 1992]). Zwicky obviously overlooked the fact that Hitler governed for a mere twelve years. Although in this period the world was indeed dramatically changed, not least by the genocide committed by the government of a civilized country, the Hitler regime represented only a part of our lives. Most of the scientists at the wind tunnel had received their degrees prior to 1933 or shortly thereafter. They had gone to perfectly good schools and had been exposed to the Weimar Republic, the resurgence of the arts and literature in the 1920s, the depression, and many other influences. All this surely made them more broad-minded than Zwicky realized. Certainly he knew nothing personal about us, a fact that is reflected in his "Report on Certain Phases of War Research in Germany."

Lasby states that much excitement arose when the Kochel site was discovered. The scientific leaders at the wind tunnels were called "impressive," and I find my name among the six listed in this context. Millikan, who would later succeed von Kármán as head of CalTech's Guggenheim Laboratory, came to a more down-to-earth conclusion after visiting Kochel. He told a group of naval officers that although we were a "very extraordinary" group of highly competent engineers and scientists, he could discover no geniuses among us. Lasby summarizes the motives of the scientists in cooperating with the Allies as "political indifference, pride, personal expediency, or some combination," ignoring the possibility that at least some of the scientists—certainly among the wind-tunnel people that I can vouch for—were opposed to Hitler and delighted with their newfound freedom.

Back at the Mittelwerk, the arrival of the Soviets was delayed beyond June 1, the date that had led everybody to the frantic work of cleaning out the plant. A systematic search was now undertaken for more Germans who had worked at Peenemünde and in the Mittelwerk and who had remained in Nordhausen. Porter, who was involved in this operation, noted that many vehicles were combing the area, each carrying a German-speaking person. Former Peenemünde people were asked whether they wished to move to the American zone in the west, and many agreed to such a move. Porter insisted that no pressure was exerted, and he did not carry a weapon. If a person was willing to go along, the whole family had to pack up in fifteen minutes and board a truck. One problem encountered during this search was that wholly unqualified people—called ringers by Porter—who wanted to use the opportunity to flee from the advancing Soviets, stepped forward and had to be checked. The trucks traveled about fifty miles southwest to Witzenhausen, a small town in the American zone.

At the time it became known that the core group around von Braun had fallen into the hands of the Americans, Josef Stalin, according to McDougall, said, "This is absolutely intolerable. We defeated the Nazi armies; we occupied Berlin and Peenemünde, but the Americans got the rocket engineers. What could be more revolting and more inexcusable? How and why was this allowed to happen?" But new developments may later have pacified Stalin. For unknown reasons, Helmut Gröttrup, one of the top people at Peenemünde, had remained in Nordhausen, and he did not sign a contract to join von Braun. Recall that he and Klaus Riedel, together with von Braun, had been arrested by the SS at Peenemünde. They were accused of concentrating on space flight rather than pushing the development of the A4 as a weapon. As an anecdotal aside, I note that Freeman observes that von Braun also attended the previously mentioned parties given by the dentist at Zinnowitz, where I had seen von Braun myself. Freeman reports that the hostess was a spy of the Gestapo, according to a 1992 interview with Gerhard Reisig, an old member of the von Braun group. Possibly Zinnowitz was where the presumably subversive remarks were made that led to the arrest. Since I had heard at Peenemünde that the dentist was a Soviet spy, the story becomes too complicated to pursue further.

After Soviet scientific personnel arrived belatedly with their army in Nordhausen, they gave Gröttrup a chance to continue his work under Soviet leadership at the Mittelwerk.[4] Gröttrup accepted the offer and assembled a group of about two hundred capable people. Most of them had originally been at Peenemünde, had moved to the Nordhausen area for the last period of the war, and had helped in the production of the V2. Among them, of course, were some of those who had turned down Porter. After the Soviet authorities took over, arrangements were made to provide food and housing, and—most important—it was promised that the group would remain in Germany. It is generally assumed that the most competent individuals from Peenemünde had joined von Braun. However, a sufficient number of capable engineers, technicians, and others stayed in the vicinity of Nordhausen to reactivate the Mittelwerk. Production of the V2 was reestablished, and about fifty missiles were newly produced or assembled from remaining parts.

The Germans' dream of remaining in their homeland came to an abrupt end on October 22, 1946, about one and a half years after the Soviet missile activity in Germany was established. On that day the men in Gröttrup's team were invited to a rousing party. In the meantime, trucks appeared at the quarters of the team members' families, who were ordered to pack up and vacate their premises. After the party, at which Gröttrup had been forewarned by a frantic telephone call from his wife, the men went home to join their families. They were forced to move by truck and railroad to the Soviet Union. They were kept aboard their train in Moscow while groups of other scientists not belonging to the rocket team were assembled on the platform. None of them could have known at the time that their lot was shared by close to twenty thousand Germans, counting the technical people and their families. About ninety railroad trains were involved in the operation. The great catch included those with skills in all areas of science and engineering. The various groups were taken to different parts of the country; some rocket engineers even lived in isolation on the island of Gorodomlia, in the Volga River between Moscow and Leningrad. Gröttrup's team still worked hard. He himself stayed in Moscow and visited his staff at the various locations, including Gorodomlia. The group produced the first successful firing of a V2 missile in the Soviet Union on October 30, 1947.

McGovern discusses the postwar work at the Mittelwerk and in the Soviet Union, referring to a number of sources in English that I did not see. The two recent books by Albring and by Magnus—scientists in Gröttrup's team—discuss the lives of the rocket designers in the Soviet Union, including the education of their children, their day-to-day life, travel, and the like. It was six or seven years before the captured scientists and their families were released to the German Democratic Republic, a state that again kept them from being free people. Knowing about these events, I am doubly happy that I did not attempt to travel to Berlin during the summer of 1945, or at any other time prior to my departure for the United States. I am sure that my father's hunch was right: I might well have been captured on a trip through the Soviet occupation zone. But neither he nor I could have known how systematically the Soviets would work to hunt down scientifically educated Germans.

How much did the German rocket experts in the Soviet Union contribute to the early launching of Sputnik and to the further successes of the Soviet space program? Albring and Magnus both provide balanced answers to this question, and much has been written about it elsewhere. The Germans undoubtedly helped to speed up the initial phases of the Soviet space and weapons programs. The later remarkable successes were largely accomplished by Russian technical teams, however. At meetings in the United States and other western countries, I had several opportunities to discuss the development of supersonic wind tunnels with Russians who worked in the field of gasdynamics. As far as I know, they did their own work, albeit with substantial assistance from the literature openly available in books and periodicals. Obviously they had many gifted contributors to the field; in fact, one of them wrote a monograph for a series of books that I edited.

In the United States, Project Paperclip—remember the clip affixed to the papers of the selected scientists—remains a topic of contentious debate even though the war ended fifty years ago and the German scientists and their families have been absorbed in varying degrees into the general population of the United States. The scientists shared this experience with about seven million Germans who entered the country before them, starting in large numbers in 1683, when thirteen Quaker families from Krefeld settled in Germantown at the northern edge of Philadelphia. Today the Paperclippers are mostly old folks in retirement, and their grandchildren are in college. But the circumstances of their entry into the United States are remarkably different from those of any other group of immigrants.

I have often thought about the ease with which I entered America in comparison with the relatively few German emigrants after 1933 who were lucky enough to receive a visa but who did not have the fame of an Albert Einstein or a Thomas Mann. I understood the objections to the whole operation raised in 1945 and later, though I often disagreed with the characterization of our group. It is important to recall that our entry into the United States coincided with the full public realization of what had gone on in the concentration camps. The facts of the mass murders caused even many military leaders to oppose the project. Again citing Lasby in the following, I find that on June 28, 1945—only about two months after the armistice in Europe—a Colonel John A. Keck held a news conference in Paris revealing the entire project. He made public the existence of a "unique war booty" represented by the "capture and interrogation of twelve hundred top-line scientists." Keck then added descriptions of fabulous future weapons that supposedly had been dreamed up by this group—weapons, I believe, that can be found only in science fiction. The well-known foreign correspondent of the *Baltimore Sun*, Philip Whitcomb, who was present on this occasion, wrote later that the United States had "no detailed plan to control [German] scientists" at the time but operated on a "day-to-day" basis. While the military thought that those picked up so far "put science ahead of nationality to volunteer their services," another large group might be busily preparing new atomic bombs and would be "unrepentant." George Fielding Eliot, a major who worked for CBS and was the military correspondent for the *New York Herald Tribune*, suggested that the "leading men of science of Germany and Japan, who have devoted their lives to contributing new methods of slaughter, ought to be confined to an island near the Antarctic circle."

The protest against Paperclip was joined by such scientific organizations as the Federation of Atomic Scientists. In Washington, I read about the view of the group in the *Bulletin of Atomic Scientists*. In contrast, I was happy to be able to become a member of the American Physical Society, the primary professional organization of physicists in the United States. With this membership, we became more broadly accepted by our professional colleagues. Lasby believes that the opposition of scientists in the United States "was fashioned almost entirely by their conviction of moral turpitude of those who worked for the cause of Hitler and the Third Reich—a conviction greatly accentuated by the mere presence of highly respected refugee scientists." Attempts were made to label the new group second rate, a correct assumption when we were compared with the leading scientists among the earlier refugees. But among these scientists I include engineers, and I believe that von Braun was one of the great engineers of our time. I am still perplexed that the views voiced about Paperclip seemed to contain the assumption that all German scientists had been, or still were, Nazis, a view that was surely far from the mark. In more recent times, certainly, nobody automatically assumed that all scientists working in the Soviet Union were confirmed Communists.

Similarly strong views on the subject were entertained in political circles. Henry Morgenthau, a close adviser to President Franklin Roosevelt, proposed prohibiting science in his well-known plan to convert postwar Germany from an industrial to an agricultural society. He obviously overlooked the fact of human curiosity. On the other hand, John J. McCloy, an assistant secretary of war who was to play a crucial role in the revival of West Germany, proposed keeping the Paperclip group permanently in the United States. The denial of the Germans' scientific and engineering expertise to other nations must have entered into this thought. Later, a high-level committee of the National Academy of Sciences said that German "scientists' concern during the war was not with politics, or with victory, but with their future professional status." The majority were "as little influenced by Nazi teaching and doctrine as any group in the population." This may be an optimistic view of the matter, but generalizations of mixed value are typical of the statements on all sides in this debate.

Finally, a note on two books that reflect the extremes of the opposing views on the subject. Bower views the Paperclip operation as a conspiracy, but it is not wholly clear from his writing who led the conspiracy. Most likely it was the military leaders who wished to acquire German scientists at all cost, including a modification or even a denial of their Nazi past. There is no question that in administering questionnaires and in conducting interrogations such as the one I went through in Virginia attempts were made to alter or rationalize memberships in Nazi organizations. This brings me back to the definition of the ambiguous term *Nazi;* however, I am reasonably sure that some who were committed followers of Hitler slipped through the system. But in discussing the wind tunnels Bower simply piles on too many errors of fact. For example, he contends that Hermann and Kurzweg were members of the SS. Supposedly these two, dressed in uniforms, held daily meetings to regale the staff with inspirational talks on ideology. Such meetings never took place during my tenure at Peenemünde and Kochel, and the same applies to the early days at the wind tunnels. In fact, I am sure that there were no SS members among the scientific staff at the Aerodynamics Institute. I never saw anybody at the wind tunnels dressed in a uniform of the numerous Nazi organizations. There are many additional errors in Bower's book related to wind-tunnel politics and individuals. For example, he identifies an innocent physicist of the wind-tunnel staff as a war criminal who later faced trial. The physicist and the war criminal had the same name, a fact that probably prompted this conclusion. During a visit to London I spoke with Bower on the telephone, and he suggested that I write to him. I did that, but he did not respond. Clearly the idea of conspiracy is too farfetched to be applied to a program supported for a variety of complex reasons by President Truman and many others in the executive branch, by Congress, and by the military, all of whom thought that the move was in the interest of the United States.

At the other end of the spectrum is Freeman. She criticizes State Department bureaucrats for assuming that all Germans brought over here were Nazis, without having ever spoken to a single one. Surely the members of the Paperclip group had gone through a great deal of checking, though the screenings may not have been wholly objective.

The supposed "collective guilt" for the crimes of the German state ought not to have kept Paperclip out of the country.

Moreover, when the search for war criminals in the United States started belatedly in the 1980s, Freeman claims that the Criminal Division of the Justice Department received questionable evidence from the Soviet Union via the Stasi, the East German secret service. She suggests that the Communists had an interest in instigating the persecution of certain individuals in the United States.

I find my observations about my own experience with Paperclip to be quite limited. I was unaware of the widely varied histories of scientists in different locations and work environments. Such situations led to widely disparate degrees of adaptation to the United States. Clarence Lasby transcribed a number of interviews that highlight the marked differences in attitude among individuals who were exposed to distinct treatment and surroundings. I must repeat that all my readings show that I was lucky to work under the navy's loose rules at a place where we wind-tunnel Germans were a tiny minority. For example, I had no unusual problems when I had to supervise American engineers and technicians at the Naval Ordnance Laboratory. At any rate, I was not alone in this situation. Lasby spoke to many Paperclip people who, like me, considered this country their home after a number of years. Several turned down offers of excellent jobs in Germany after that country's reconstruction, and I had the same experience. In spite of regular visits to Germany, sabbaticals of teaching at German universities, and contact with family and young postwar friends, my home is in the United States.

Again following Lasby, during the main migration—between May 1945 and December 1952—624 scientists arrived in the United States. This number includes a few Austrians. By 1960 the volume *American Men of Science* listed 126 individuals from this group. Of the total, only about 20 percent were engaged in missile work, including space applications. A small number of Paperclip people returned to their homeland, but 80 to 85 percent of those remaining became American citizens. Lasby points to the surprisingly high number of papers, books, and patents produced by the group.

The *Central Falls Victory* had carried me to the dock in Manhattan in January 1946. Since that time a few members of the Paperclip group have returned to their homeland, and some have died, but most live in retirement throughout the United States. Many of their grandchildren have now finished college. I am saddened by the recurrent negative views published about this diverse group. For example, Hunt, as late as 1991, still employed the cliché "Nazi scientists" in the title of her Paperclip book. To the best of my knowledge, none of the Paperclip scientists engaged in sabotage, spying, forming Nazi-like organizations, or other criminal activity, as had been expected by many of the opponents of the program. I believe it is fair to say that group members did not take jobs from native-born Americans, and significant scientific and technical contributions were offered to society by the immigrants, who entered the country under such unusual circumstances.

The last—and most difficult—topic of my backward look concerns the Mittelwerk, the site of the mass production of the V2. I had seen this place with Günther Hermann in the spring of 1945 (see chapter 8), and the event is still vivid in my memory. Our visit shortly before the end of the war to retrieve the original wind-tunnel reports was as short as we could make it. But it was long enough to enable us to understand to some extent what was going on in this place. We found that the production of the V2 was being carried out to a large extent by inmates of a concentration camp. Strangely, I had not been forewarned by anybody about what to expect, and possibly nobody at Kochel knew anything about the events in the Kohnstein. But at least Rudolf Hermann, who was in frequent contact with the leaders at Peenemünde, might have been informed. He said nothing prior to our departure to pick up the reports, however. Beyond the initial discovery of the concentration camps by American troops and the gathering of the remaining V2 missiles, it has taken historians years to penetrate the actual operation of the Mittelwerk. Even now are new facts coming to light about the production of the missile between 1943 and 1945.

Here is a short history of the operation based on the sources cited in the notes, discussions and correspondence with the historians named in the acknowledgments, and my visits to the site.[5] The mining of anhydrite ($CaSO_4$)—a substance of importance to the building industry to produce gypsum and plaster of Paris—was started in 1917 by a private company. Horizontal tunnels were driven into the Kohnstein, and open-pit mining took place on one side of the mountain. In 1935, two years after Hitler came to power, the Interior Ministry requisitioned the by-now extensive tunnels in the mountain. A government-controlled company was founded, and additional blasting and digging were undertaken to extend and enlarge the tunnel system. An anhydrite mine lends itself well to such purposes, because the natural material is as solid as rock, so little internal support is required. Starting with the existing underground maze, two parallel main tunnels about 1.8 kilometers (a little over a mile) in length and several domed caverns up to 80 meters (260 feet) high were completed. As part of the preparation for war, the caves were to be used for the secret storage of such raw materials as oil and chemicals. In August 1943 it was decided to switch the use of the tunnels to the mass production of the V1, V2, and other military devices. Consequently, the initially planned fabrication of the V2 at Peenemünde was to be abandoned. I must refer to the cited literature to follow the Byzantine dealings between the SS, Peenemünde officials, and private contractors that preceded the founding of the Mittelwerk. At this stage of the fight for control of rocketry, the SS had wrested authority over the V2 from army ordnance, and Dornberger was practically shut out of the decision-making process. Himmler appointed the previously mentioned SS general Hans Kammler, who had experience with the construction of concentration camps, to take over at the Mittelwerk. The first group of inmates, from Buchenwald, began working on further extension of the tunnel network, and by September the number of inmates had increased to about three thousand, with about twelve thousand prisoners present in February 1944.

Since no housing was available on site, many prisoners had to live inside the Kohnstein while performing the extremely hard work of building a third extension to the caverns. The underground life was beyond the strength of many; marginal food supplies, dust from the blasting, and the lack of sleeping accommodations, ventilation, and sanitary facilities contributed to creating an inhumane environment and causing a high death rate. It is estimated that close to three thousand prisoners perished in the caverns during the winter of 1943–1944. But the work was brutally pushed, and in the end the underground plant covered about 1.3 million square feet (120,000 square meters).

Close to the Kohnstein, a new concentration camp called Dora was constructed as an offshoot of the Buchenwald camp near Weimar, the city of Goethe and Schiller. Gradually prisoners living in the mountain moved to Dora, and between the twelve-hour shifts they walked the short distance to and from work. In 1945, I saw a change of these shifts, with a column of prisoners in their flimsy striped suits emerging from the huge main portal of the Mittelwerk. An SS physician who testified in 1947 at the Nordhausen-Dora war crimes trial held in Dachau, the site of a concentration camp near Munich, stated that the inmates were in a bad state of malnutrition. Infections, in particular on the legs, were worse than any he had seen. But he was told that, given the aims at the Mittelwerk, it was unimportant how many workers would lose their lives.

The inmate force was bolstered by technicians and scientists who had been arrested in several countries occupied by Germany, in particular

France and Poland. Even after all the prisoners had moved from the plant to the camp, the treatment inside the plant remained terrible. Beatings and other physical punishments were common. As we had heard during our 1945 visit, infractions of discipline and attempts at sabotage were occasionally punished by hangings from wire loops attached to the cranes installed in the high ceilings. All this kept the crematory at Dora going; that facility can still be seen today. Dora became independent from Buchenwald, and other camps, such as Ellrich, were attached to it.

Toward the end of the war large numbers of additional prisoners arrived from concentration camps in the east, situated in the path of the advancing Red Army. The final blows to the surviving prisoners were dealt during the last days of the war. During the first week of April 1945, inmates of Dora and the associated complex of camps were piled into trains of open freight cars, without food or water. The trains rolled north to escape a meeting with the Allied army. Dora itself at this time contained seventeen thousand prisoners guarded by about one thousand SS men. Many of the prisoners died in the freight cars. Those who could not be fit into the limited space were forced to walk. Many collapsed on the way or were shot by the guards. In the most extreme case of mass murder during this death march, 1,016 prisoners were burned to death in a large barn near the town of Gardelegen. The mayor of the city arranged for burial in a mass grave, an action that allowed for the counting of those killed.

It is difficult to determine the exact number of prisoners who worked at the Mittelwerk before the Americans arrived on April 11, 1945. It has been conservatively estimated that sixty thousand inmates of concentration camps and other prisoners were involved in the Mittelwerk operation from September 1943 until the end of the war. Of these, it is now thought, about twenty thousand died. Therefore, about one-third lost their lives in or near the caverns of the Kohnstein while producing the V2 and other weapons. The total control by the SS, the high death rate of inmates from malnutrition, weakness, and diseases, and the number of prisoners that were murdered by the SS guards probably had no parallel. Although many factories in Germany during the war made use of forced labor from several sources, with some of the major plants using inmates of concentration camps in their labor force, the Mittelwerk may well have represented the extreme case of this kind of operation.

Aside from the prisoners, a large number of German engineers and technicians worked on the V2. The SS used an iron fist to maintain discipline, a discipline that extended to the German staff, which was completely excluded from proposing changes in the treatment of the prisoners in the interest of the quality of production. Germans were not even permitted to talk to prisoners other than on technical matters relating to production. Any hint about easing this or that action in the interest of the prisoners led to immediate threats by the SS, who would suggest that the questioner could quickly be ordered to join the prisoners. Such threats had to be taken seriously. Most men are not heroes, a fact that some of the younger postwar historians do not seem to consider in their censure of those caught in extreme circumstances, like those existing in the Mittelwerk.

How many German specialists were involved in this macabre endeavor? A company to run the complex was founded in September 1943. Several German industrialists served on the board, and a consulting group was formed. Dornberger served with this latter group, and so did Kammler, the real boss of the Mittelwerk. The organizational charts changed as time went on. Arthur Rudolph—who had run the fabrication of sample V2s at Peenemünde—moved with a few associates in September 1943 to Nordhausen, where he assumed one of the two most important jobs for the production of the V2. He was the director for the "assembly, control, energy, civil and prisoner deployment." The first lot of fifty-two V2s left the plant in January 1944, a month in which 679 inmates died. In July 1944 the number of people working on the V2 reached its height. Of the roughly 12,000 inmates of the concentration camps at this time, about 5,000 were involved with the V2. To this we must add 3,400 German civilians, of whom 500 were in supervisory positions, 1,000 were in administration, 1,500 were skilled laborers, and 400 did general work. The total V2 work force—prisoners and civilians—had reached its highest number of 8,400. Not all the Germans actually worked in the Kohnstein, of course. Offices had been opened in buildings in the vicinity, and the bureaucracy functioned up to the last minute of the war. This fact makes the reading of correspondence, orders, and the like particularly eerie. Such matters as death rates of prisoners, rail movements of new arrivals, and storage of turbopumps are treated as if the staff were running a chicken farm. Strangely, in March 1945, the number of prisoners working on the V2 had shrunk to about 2,000, while the civilian group had grown to 4,900, for a total of 6,900, reversing the numerical relation between prisoners and civilians. In this discussion I have used the word *German* for the civilian specialists, who came largely from Peenemünde, though many of the prisoners were also German citizens.

Close to six thousand V2 missiles were produced in the Mittelwerk, starting in fall 1943 and ending in April 1945. Many of these missiles failed, many were used for testing, and some were stored. Only a little over half of all the missiles produced were actually fired in the field, and a majority of them were aimed at such continental targets as Antwerp rather than at England. It is clear from these numbers that as a weapon the V2 must be called a failure. Here I disregard the fact that the V2, like the V1, produced a strong psychological effect in England late in the war, when German air raids had essentially ceased and new weapons that could reach London were not expected. The fact remains that the total explosive power of all V2 warheads fired in the field was roughly equal to that of the bombs dropped during a single major air raid on a German city.

What happened to those who were responsible for production at the Mittelwerk? The fate of Kammler is unknown. After the failed assassination attempt in July 1944, Hitler installed Himmler as the commander of the Wehrmacht within Germany and as head of weapons production for the army. Kammler was given responsibility by Himmler for all missile production, including that of the V2. In the last days of the war he returned to the service as an SS general to command an army division. It is known that in April 1945 he ordered the shooting of 208 Russian workers, a group that included 77 women (see Eisfeld). It appears that he was in Prague at the end of the war. Various accounts suggest that he may have committed suicide, been killed in fighting, or even been shot by an associate.

For someone like me, who had only a brief encounter with the Mittelwerk at a time when I did not know the facts that I have mentioned here, it is difficult to understand how anyone from Peenemünde or from companies involved in other projects could have maintained his sanity given the conditions in the Kohnstein. To even begin to understand the environment in this place, it is important to read the descriptions by surviving prisoners, which are listed in the cited literature. It is even more difficult—if not impossible—to comprehend the personal suffering that lies behind the tragic statistics I have given.[6]

It is, of course, impossible to weigh the degrees of involvement in criminal actions by the men from Peenemünde who were transplanted to the Mittelwerk. The last thing I have a right to do is take a moral stance about a situation in which I had no part. However, I shall record some of the facts of at least one career: that of Arthur Rudolph, whom I had once met briefly at the Baltic. He was one of the most important people at the Mittelwerk, and in the United States he carried the responsibility for the development of the Saturn V, the rocket that carried men to the moon in July 1969. I have no doubt that this first expedition to another heavenly body will go down in history as one of the most important human achievements of the waning twentieth century. Rudolph was received by presidents and was honored with NASA's highest decoration, the Distinguished Service Medal. He retired in 1969, after the moon shot. But what about his earlier career?

Rudolph joined the Nazi party in 1931, two years before Hitler came to power. From a written deposition given by him in May 1947—in conjunction with the Dachau trial—Eisfeld quotes: "After careful thought, I decided to join the National Socialist party, a legal organization, to contribute to the perpetuation of Western culture. In the following years, this decision did not prove to be a mistake from a business viewpoint" (my translation). The early party membership and the statement at the Dachau trial clearly fit my definition of a Nazi. Although he also seems to have been recognized as a "100 per cent Nazi" by Project Overcast interrogators, Rudolph was permitted to join von Braun's core group. In a second statement in the deposition Rudolph acknowledged that virtually all who worked in the Mittelwerk were aware of the prevailing conditions, including the garroting of prisoners who were hanged from the cranes. It was later noted that prisoners but not the civilian staff were forced to witness the hangings. It was said that Rudolph voluntarily witnessed executions. Additional depositions by von Braun and others from his group in the United States were given at the same trial, but no action was taken against anyone at the time in response to these testimonies.

Nothing much happened until Congress asked the Department of Justice to determine whether war criminals had found shelter in the United States, and the Office of Special Investigations (OSI) was formed. In October 1982, long after his retirement, Rudolph was interrogated by OSI representatives in San Jose, California. It appears that his recollections of the past had faded. But documents had come to light noting that army investigators shortly after the war had classified Rudolph as a war criminal. At the time of the OSI investigation he was accused because of his recurrent insistence that the supply of prisoners from concentration camps be replenished. Such requests led to the certain death of a number of them, a fact of which he must have been aware. Furthermore, the SS leadership reported complaints by SS physicians at the camp Dora that the Peenemünde work force on occasion physically attacked prisoners. The directors of the Mittelwerk were advised by their civilian bosses that this behavior was to stop. It is not clear why such mistreatment was not noticed by those at the site or why Rudolph did not put a stop to it in his own area. Some of the surviving Dora prisoners said that they thought Rudolph had participated in selections for the hangings, but I have seen no proof of this. Other prisoners praised him, stating that whenever Rudolph made the rounds of the plant, conditions improved. Rudolph contended that he had always fought the SS to better the conditions of the prisoners in the interest of efficiency. He claimed that he was consistently rebuffed and threatened with arrest, and again there is no reason not to believe him. Based on these findings, the OSI proposed stripping him of his U.S. citizenship. To forestall this event Rudolph voluntarily gave up his citizenship and left the country with his wife to settle in Hamburg, Germany. Freeman, a staunch defender of Rudolph, claims that "from its inception the osi had largely been a joint venture with Moscow." Again she asserts that the East Germans supplied material to accuse possible war criminals.

Rudolph now began a long and unsuccessful fight for rehabilitation. A German court, after investigating the evidence, including an accusation of one murder, restored Rudolph's German citizenship in 1987, citing insufficient evidence to convict him of war crimes. Humanitarian notions may have played a part in this decision, since many American and German friends and colleagues testified in defense of his character. In fact, a group calling itself Friends of Arthur Rudolph unsuccessfully petitioned Senator Howell Heflin of Alabama to ask the Senate Judiciary Committee to intercede in his behalf. His defenders did not deny the basic facts about the Mittelwerk, but they felt that not enough consideration had been given to his major contributions to the U.S. space program. In 1991, Rudolph attempted unsuccessfully to regain his U.S. citizenship, and in 1993 a federal judge in California dismissed a lawsuit filed by him. His application for an immigration visa was turned down by Canada's Federal Court of Appeal on May 1, 1992. The reason given in the document is as follows: "The applicant, as production director, admittedly called for, made use of, and directed forced labor by foreign prisoners in the production of the V-2 rocket at Mittelwerk in the years 1943–1945. The conditions under which the prisoners worked were indescribably brutal."

Much has been written for and against Rudolph. It seems that the key to fairly judging him lies in his relation to the prisoners and his attitude toward them. The Mittelwerk records demonstrate that Rudolph kept asking for more prisoners to replenish the ranks of his workers. This is true, although the "civilian" work force steadily increased. But, more surprisingly, the records show that Rudolph, without any SS pressure behind him, was the first person to propose the use of concentration camp inmates at Peenemünde for the assembly of A4s. During a visit to the Heinkel-Werke (an aircraft plant) at Oranienburg, a small town near Berlin, he was informed by the management that, owing to the frequent turnover of foreign contract labor from France, Poland, and other countries, production suffered. Oranienburg was also the site of a concentration camp (now transformed into a frightening museum, which I visited in 1994). Trying to solve their labor problems, managers of the Heinkel-Werke asked for three hundred inmates from the concentration camp to add to the work force. The SS agreed, and after some introductory schooling, these prisoners proved to be good workers. Rudolph was impressed by this solution to the labor problem, and he wrote a long memorandum dated April 16, 1943, on his trip to Heinkel for distribution to some of the Peenemünde authorities (not including von Braun). In the last paragraph of the memorandum he proposed adopting this procedure in his department at Peenemünde and suggested consulting the SS. He saw a particular advantage in the fact that the SS took over the complete handling of the prisoners by providing their selection, housing, food, and so forth, permitting him to concentrate exclusively on the technical side of the operation.

It is essential in judging Rudolph to note the timing of his proposal. The events described occurred in April 1943, four months before the major air raid that I have described, and prior to the decision to move the V2 production to the caverns in the Kohnstein. Undoubtedly, no such proposal needed to be made in the context of Peenemünde. In fact, Rudolph would not have compromised himself had he not reacted to his experience at the Heinkel-Werke at all.

This early, quite voluntarily advanced, spontaneous idea of using prisoners, followed by the Mittelwerk action, sets Rudolph clearly apart from most others involved in the final mass production of the missile. Since he had stated at the Dachau trial that everybody knew what was going on in the Mittelwerk, and since he obviously knew why his work force kept decreasing, his actions were tantamount to selecting a certain number of inmates to die. No one else—to my knowledge—had suggested contacting the SS for a labor supply at Peenemünde. In fact, Dornberger and von Braun at roughly the same time had made every conceivable effort to keep the SS away from the A4 development and manufacture. In sum, Rudolph's actions at the Mittelwerk demonstrate his personal guilt. His fight to clear his name ended with his death in December 1995.

Now to a few final personal remarks about von Braun. I am, of course, aware that my views on the character of von Braun contradict those of most recent authors. For this reason, I have to explain myself. To start, I can attest categorically to the fact that von Braun was no Nazi. Again, this may seem strange in view of his 1937 party membership and his honorific SS officer rank. However, his personality, his family background, and the spirit of his boarding school—a school that followed the ideas of the great educator Hermann Lietz, just like my school—all speak against his falling for an ideology such as Hitler's. In addition, at this time one developed a sixth sense with which to detect a person's attitude about the regime when engaged in private conversation on almost any subject. The discussions I had with von Braun told me that his outlook on life was far removed from the prevailing ideology.

This leaves us with his career at Peenemünde and his relations with the Mittelwerk. The work at Peenemünde led him to see a future transition to space flight that he thought could become a reality only in the United States after the war. During my time at Peenemünde, however, I did not hear von Braun speak of space flight. He surely knew that Peenemünde existed solely to build a weapon, the A4 missile. In contrast, it is amazing that the view of Peenemünde as the first space flight center could be discussed seriously in much of the postwar literature. A renewed expression of this myth is found in Freeman's 1993 book, and it is repeated in its German translation as late as 1995. An old Peenemünde hand wrote an introduction to the books attacking the media. He is incensed that Peenemünde is still called a *Waffenschmiede* (weapons smithy) and that it is said falsely that the army used concentration camp inmates as laborers.

At a Peenemünde meeting that included von Braun held on August 25, 1943—four months after Rudolph's call for concentration camp labor—the distribution of work among several subcontractors was discussed. One company was using caves near Saarbrücken to make parts for the A4 with the aid of inmates of a nearby concentration camp. But what about the Mittelwerk? In von Braun's written deposition sent to the Nordhausen-Dora war crimes trial, he acknowledged that he was at the Mittelwerk fifteen to twenty times to discuss problems related to technical changes in the missile. He was quickly ushered into and out of an underground conference room, and he must also have been there in the early days of the Mittelwerk, when the inmates of Dora still had to live in the extreme squalor of the tunnels. These visits were sufficient to let him see what was going on. Although talking about the plant was forbidden, he related his impressions to close friends, including Ernst Stuhlinger, after his first visit. According to Stuhlinger and Ordway's biography, he said, "It is hellish. My spontaneous reaction was to talk to one of the SS guards, only to be told with unmistakable harshness that I should mind my own business, or find myself in the same striped fatigues! I would never have believed that human beings can sink that low, but I realized that any attempt of reasoning on humane grounds would be utterly futile. Such arguments as decency, fairness, morality, or ethics did not count here. These individuals had drifted so far away from even the most basic principles of human morality that this scene of gigantic suffering left them entirely untouched."

I further find that Dornberger and von Braun said they spent hours trying to figure out how they might help. Later von Braun said, "We did not arrive at any good idea. Even if both us had resigned—a step that would have converted us into inmates immediately—we would not have helped the prisoners. Hitler and the SS commanders, in their frenzied determination not to give up their mad race toward doom, would certainly continue all the Mittelwerk operations—V2 rockets, airplane engines, antiaircraft missiles, submarine components, fighter planes. Inmates and civilian workers would have been driven only more relentlessly." Knowing von Braun and Stuhlinger, I simply believe these statements to be true.

After the war, von Braun expressed his views on the terrible conditions under which the V2 was manufactured. Former prisoners at Dora—for example, two French inmates who had written books on the subject—were incensed that von Braun could have tolerated the conditions at the Mittelwerk. Newspapers and magazines took him to task in the 1960s, and he had a tough time defending his actions. Indeed, there was little he could say other than stating that he had tried his best and failed to improve the conditions under which the inmates of the concentration camps labored in the Mittelwerk.

Surely von Braun was nearly obsessed with the development of long-range missiles, an obsession that originated in the space-flight dreams of his early years. To repeat, the orders given by Himmler via Kammler—to construct the V2 in an underground plant manned largely by helpless inmates from concentration camps—caught von Braun in a dreadful trap. We will not readily find another man in our time who on the one hand has been admired like few others—he received the major decorations this country can offer, he was praised by presidents, he was revered by the young, he became a hero in the German Bundesrepublik—but on the other hand has been called a mass murderer, an inhuman technocrat, an amoral person, an engineer who would have worked for the devil to achieve his aims. This sharp exchange of views about Wernher von Braun continues. I note sadly that those who assail von Braun's character have not proposed what he ought to have done. He was a good pilot, had an airplane at his disposal, and could have flown to Switzerland. He could have resigned, leaving his life's work and his associates of many years, and he could have spoken with Hitler and Himmler. All but fleeing Germany would have produced the certain result of incarceration in a concentration camp.

Michael J. Neufeld's 1995 book *The Rocket and the Reich,* dealing with the period of A4 development in Germany, is the best history of the subject that I have seen. Neufeld stands out by providing astounding detail gleaned from archives and interviews. Moreover, his clear and straightforward description of the technical side of the A4 development for once includes the wind tunnels. In fact, Neufeld lists aerodynamics as one of the essential ingredients of the development of the A4. While he has high regard for von Braun's leadership and for the technical achievements of the Peenemünde staff, he severely condemns von Braun on moral grounds. More recent archival discoveries point to a broader involvement of von Braun with the planning for and use of concentration camp inmates as workers in the Mittelwerk. For one, a memorandum dated August 25, 1943, four months after

Rudolph's initial proposal, shows that von Braun was one of seven participants at the Peenemünde meeting on A4 production. As noted above, the distribution of the work among several firms was at stake. Aside from considerations of competence, the location of the firms in relation to possible air attacks had to be considered. The caves near the city of Saarbrücken were discussed, and in this connection it was noted that workers had been recruited from a concentration camp.

Moreover, a letter of August 15, 1944, from von Braun to A. Sawatzki, the head of planning for the Mittelwerk, relates the fact that von Braun and Eberhard Rees, my Haus 30 associate, visited Buchenwald to select prisoners suitable for work on the V2 assembly in the Mittelwerk. During this visit von Braun dealt with, among others, Pister, the SS commandant of the camp, who was sentenced to death in 1947. Near the end of this important letter discovered by Neufeld, von Braun pleads for preferential treatment of a selected French professor of physics. Here von Braun may well have hoped to improve the life of some of the prisoners by moving them to the Mittelwerk. It appears to me that within the macabre hierarchy of concentration camps, such a move might indeed have improved a prisoner's survival chances. Here I go back to my belief that unless a person enmeshed in this dreadful environment committed a personal crime, we must be careful in our retroactive judgment. My short visit at the Mittelwerk provided some insight into the conditions there, and that insight has not left me. I do not know what I would have done if I had been asked to work there.

There can be no question about the facts concerning the Mittelwerk, based on what I saw there before the end of the war and on the literature cited here. If it comes to judging the varied involvement of individuals, however, I believe there remains the problem of interpreting the many archival sources that have come to light concerning Peenemünde. True to German tradition, letters, memorandums, and orders were exchanged among the many groups involved with the V2 and Wasserfall. This exchange did not cease until the whole country was occupied in 1945. Interpreting such material in retrospect is difficult. Struggles between individuals, between the army and the SS, the air force and the army, and so on played a role in the exchanges. Historians tell me that careful interpretation of archival data is in fact possible, but having been with the Aerodynamics Institute of the German army during the last two years of the war—albeit in a minor niche of the Peenemünde action—I often have doubts about how the many pieces of the puzzle presented in the archives fit together to express reality.[7] At any rate, I have not seen documents that prove von Braun responsible for a crime.

The armistice in Europe is now fifty years behind us, but the subject of

Peenemünde will undoubtedly continue to be of interest. My single brief visit to the Mittelwerk led me to study some of the history of the place, and a small fraction of the material I have gathered about it appears here. The brutal conditions in the Mittelwerk—as has on occasion been said before—turned the V2 into a weapon in whose production many more people were killed than were killed in its use against the enemy.

How is this part of the history of the war viewed in Germany? In older books about Peenemünde, the underlying thoughts were often "from Peenemünde to the moon" or "the detour via the weapon." The first book by someone closely associated with Peenemünde that deals in detail with the Mittelwerk and related problems is that by Stuhlinger and Ordway. The former German Democratic Republic preserved the Mittelwerk and set up the memorial site at the Kohnstein. It will still take time for the facts of V2 production to become generally known as part of the history of the Hitler period. The small museum at Peenemünde contains an exhibit that demonstrates some understanding of the subject. I visited this museum on October 13, 1992, a year in which festivities were planned at Peenemünde to commemorate a fifty-year jubilee of the first successful firing of the A4 on October 3, 1942. It took outraged comments by major newspapers in Britain and Germany before the governor of Pomerania and the ministry in Bonn responsible for space exploration, which were sponsoring the event, decided to cancel the ceremonies. I listened to an undersecretary in Bonn who appeared on television and called the uproar wholly incomprehensible.

Officials like him are either ignorant about the circumstances in which the production of this miracle weapon was entangled or they are most insensitive. At roughly the same time, a statue of Air Marshal Arthur "Bomber" Harris was ceremoniously unveiled in London. Surely the devastation and the number of deaths caused by the Allied airplanes vastly exceed that wrought by the V1 or the V2. However, a comparison of this event with the celebration of the first successful shot of the A4 by Bonn officials ignores the catastrophic conditions under which the V-weapons were built.

Although the Mittelwerk was far from Peenemünde, its long shadow was cast over all the efforts in the technical development of the first guided missile, the A4. This includes the pioneering work on supersonic aerodynamics. The memorial site of the Mittelwerk, now concerned primarily with preservation and history, has become a landmark. This designation guarantees the permanence of its existence. In Germany, I heard of plans to extend the memorial in Nordhausen to a more ambitious museum. A part of the underground plant could be opened, and the small collection of art by the former prisoners could be augmented by historical exhibits. All this would add another memorial to the few locations in Germany where the terror of the Hitler period is remembered. My brief visit to the Mittelwerk to retrieve the original documentation of the work on supersonic aerodynamics confronted me with the dark side of the development of the A4. This dramatic experience changed my early views of my stay at Peenemünde and became an essential part of my recollections, which I now close.

Chronology of Peenemünde and the Wind Tunnels

1929 The ballistics department at army headquarters in Berlin, under the leadership of Professor Karl Becker—who was also a general—considers the value of rocketry to augment artillery.

1930 Walter Dornberger (1895–1980), an engineer and a captain in the army, joins the study. Later, as a colonel and then a general, he becomes the military head of the technical side of the army laboratories at Peenemünde.

1932 The army starts to experiment with rockets at Kummersdorf West, an artillery testing ground near Berlin. Dornberger appoints the nineteen-year-old student Wernher von Braun (1912–1977) to assist.

1933 In January, Hitler comes to power. The A1, a liquid-propellant rocket with sadly deficient aerodynamic characteristics, is developed. The various rockets are called *Aggregat,* a word that is nearly identical in English, leading to the abbreviations A1, A2, etc., with the A4 later designated as V2.

1934 Primarily under the leadership of von Braun, the A2 is developed. In December the missile rises more than 2,000 meters (6,600 feet), but few details of the A2 are recorded. Dornberger subsequently notes that from the beginning von Braun was the driving force of the team. Rudolf Hermann (1905–1991) works at Wieselsberger's institute at Aachen. They are building a 10-by-10-centimeter supersonic wind tunnel operating to M=3.

1935 Development of the A3, a much advanced device, begins.

1936 Further testing of rockets in the densely populated area near Berlin becomes impossible. Consequently, a *Heeresversuchsanstalt* (army experimental laboratory), with Wernher von Braun as director, is proposed; it would be located at Peenemünde, a Baltic fishing village on the island of Usedom. Eventually more than 10,000 people work at the rocket laboratories. First measurements of the aerodynamic characteristics of the A3 are performed at

Aachen. Rudolf Hermann begins the design of the 40-by-40-centimeter (16-by-16-inch) supersonic wind tunnel destined for Peenemünde.

1937 In December an A3 is launched from the small island of Oie in the Baltic close to the mainland and near the laboratories. This offshore firing site is chosen because the early rockets often developed erratic trajectories. The Aerodynamics Institute is founded; Hermann, the director, has a staff of sixteen.

1939 The development of the A4 is under way. The first 40-by-40-centimeter wind tunnel begins operation. A balance to measure aerodynamic characteristics, such as the drag and lift, of models in the wind tunnel becomes available, permitting the determination of drag and lift of the A4. The wind-tunnel staff now numbers sixty.

1940 First test-stand firing of a liquid-propellant rocket motor of 24 tons (52,800 pounds) of thrust force. This motor will propel the A4. Designs for a two-stage intercontinental rocket are started.

1941 On August 20, Hitler approves the development of the A4. Plans are laid for an antiaircraft missile, the Wasserfall.

1942 On October 3 the first successful firing of the A4 takes place. On a trajectory over the Baltic Sea parallel to the coast, a range of 191 kilometers (119 miles) and an altitude of 85 kilometers (53 miles) are achieved in a five-minute flight. A maximum speed of 1,340 meters per second (3,000 mph) is recorded. This year also sees the first successful firing of solid-propellant rockets from a submerged submarine. Göring, the head of the German air force, authorizes the development of an antiaircraft missile, a project that can be carried out only by the army.

1943 Hitler gives top priority to A4 development. The wind-tunnel staff has increased to two hundred. Late in May the author joins the staff. Mach number 4.4 is achieved in the wind tunnel. During the night of August 17–18, a British air raid destroys much of the army laboratory. Little damage is done to the outlying rocket test stands, however, and the wind tunnels are not touched. A decision is made to move the entire Aerodynamics Institute, which now has about two hundred employees, to the small town of Kochel in Bavaria, the site that had been chosen for a large hypersonic wind tunnel. Work to receive the Peenemünde tunnels starts at Kochel. It is said that the air raid barely delays the previously initiated plan for mass production of the A4 in the Mittelwerk, an underground factory in the middle of Germany (see chapters 8 and 11). The training of special artillery units begins, to teach the artillerists to fire the A4 in the field. On November 5, Rocket Battery 444 starts actual practice firing in Poland. Practice firing at other sites follows.

1944 In January the Kochel installation is formally incorporated under the name of the Wasserbau Versuchsanstalt (wva). (It is difficult to translate this name. *Wasserbau* deals with the aspect of civil engineering concerned with dams, weirs, and the like. *Versuchsanstalt* translates roughly as testing labora-tory.) The first successful launching of the Wasserfall takes place on February 29. Production of the A4 rises to nine hundred per month by the end of the year. The total production in 1944 numbers above 4,000, built at a cost of 40,000 German marks per unit. All in all, about 6,000 missiles are produced until April 1945. On September 5 and 6, 1944, the first A4 firings take place at the front. Remarkably, the very first shot is aimed at Paris, with Liége being next in line. Firing at London begins on September 8. From now on things become complicated. (I checked V2 firing data in many of the books cited in the notes and found surprisingly discordant figures on the subject. It appears that the production numbers given are about right. Between 3,000 and 3,500 V2 rockets were actually fired in the field. Of this number, 1,000 to 1,300 were shot toward England, and the rest were aimed at continental locations, in particular at the port of Antwerp, which was most important to the Allies. An unknown number of rockets malfunctioned, and continuous testing and prac-tice firing by the soldiers in Peenemünde and Poland, respectively, required many additional missiles. By now the A4 was, of course, the V2, and about 2,700 deaths and 6,300 injuries were caused by the weapon in England. In-cluding continental victims, up to 5,000 people were killed overall. Compar-ing these numbers with a similar count for the V1, the flying bomb, we find the latter weapon to be more effective. Here I may be on safer ground, numer-ically speaking. From the first launch of a V1 against England on June 13 until September 1944, about 5,400 V1s reached England. In this period more than 6,000 people were killed by the V1 and about 17,000 injured. Considering that several effective countermeasures could be employed against the V1—barrage balloons, antiaircraft guns, fighter planes to shoot down the V1 or to tip its wings to make it crash—the flying bomb beats the missile as a German wonder weapon of World War II by any standard. In these considerations, the much lower cost of the V1 is not even included.)

1945   The last V2 of the war reaches England on March 27, 1945, with the last flying bomb, the V1, following two days later. Testing of the V2 to discover further causes of the malfunctions proceeds without interruption at Peene-münde, with the last firing taking place on February 14. Three days after this, the evacuation of Peenemünde begins. Many of the employees drafted from nearby villages and towns simply go home. The others pack their belongings and leave by truck. (I do not know what happened to the inhabitants of Tras-senheide, the camp of prisoners and forced laborers.) The Red Army occupies Peenemünde in March. The core group of about 130 staff members, including von Braun and Dornberger, stays together, traveling to Bavaria. Contact is made with units of the American army on May 8. Kochel is exposed to brief artillery fire by American armored units. The wva gets through the last days of the war without any losses of personnel or equipment, however. The in-stitution is taken over by the American army, and scientific teams begin to arrive a few days later.

~Notes

Chapter 1: From the Russian Front to Peenemünde

1. Albert Defant's lifework is summarized in his two-volume treatise *Physical Oceanography* (Oxford: Pergamon Press, 1961).

2. Peter P. Wegener, "Die hydrographischen Ergebnisse der deutschen Spitzbergen Expedition 1938" (The hydrographic results of the German Spitsbergen expedition, 1938), Friedrich-Wilhelms-Universität, Berlin, May 11, 1943.

3. Martin Schwarzbach, *Alfred Wegener und die Drift der Kontinente* (Alfred Wegener and the drift of the continents), Grosse Naturforscher, vol. 42 (Stuttgart: Wissenschaftliche Verlagsgesellschaft MBH, 1980). See also Martin Schwarzbach, *Alfred Wegener: The Father of Continental Drift*, addenda to the English edition by Anthony Hallam and I. Bernard Cohen (Madison, Wisc.: Science Tech, 1986). Both Alfred and Kurt strongly influenced my choice of geophysics as a subject of study and my interest in the Arctic, without ever having suggested these topics. I knew Kurt much better, because Alfred died on the ice cap of Greenland on November 1, 1930, shortly after his fiftieth birthday, when I was only thirteen years old. Alfred first crossed the ice cap in 1913 with the Danish explorer Johan Peter Koch. After making various studies of Greenland, he led the famous 1930–1931 expedition, in which scientific stations were established on the two coasts and in the center of the ice cap. Geological and geophysical experiments were performed at the coastal stations as part of a search for evidence of the validity of the drift hypothesis. Alfred perished on the return from the central station when exceptionally bad weather set in. After his death, Kurt took over the direction of the work in Greenland and the publication of the results. For further information on the expeditions, see Else Wegener, *Alfred Wegener's letzte Grönlandfahrt* (Alfred Wegener's last Greenland voyage), 7th ed. (Leipzig: F. A. Brockhaus, 1936).

4. *Supersonic* implies a motion that is faster than the speed of sound, the speed with which the sound of a drumbeat, of a passing train, or of speech approaches you. The speed of sound in such gases as air depends on the temperature, with higher speeds attained at higher temperatures. At an average sea-level temperature, the speed of sound in air is about 340 meters per second, or 1,100 feet per second. An

aircraft flying at that speed moves at about 1,200 kilometers per hour, or 760 miles per hour. A missile, a rifle bullet, or an aircraft such as the Concorde moves through the air at speeds that exceed the speed of propagation of sound. As applied to a wind tunnel, supersonic means that the airstream in which models are tested moves faster than the speed of sound.

5. Walter Dornberger (1895–1980), *Peenemünde: Die Geschichte der V-Waffen* (Peenemünde: The history of the V-weapons), 3rd ed. (Esslingen: Bechtle Verlag, 1981). This book is an extended and revised edition of Dornberger's 1952 book *V2— Der Schuss ins Weltall* (V2—The shot into space), also published by Bechtle. The 1981 book has an introduction by Eberhard Rees, whom we shall encounter in chapter 4. Dornberger's role as the head of the Peenemünde establishment will be detailed in chapters 4 and 5. As a young officer in the 1920s he foresaw the application of rockets to warfare and to space flight. The 1952 edition of Dornberger's book also appeared in English: *V-2*, trans. James Cleugh and Geoffrey Halliday (New York: Viking, 1954).

Chapter 2: First Impressions

1. Rudolf Hermann (1905–1991) was well chosen by Wernher von Braun as the first and only director of the Aerodynamics Institute. Hermann held a doctorate in fluid dynamics; he had also received the degree of Dr. habil. at the Technical University (Technische Hochschule) of Aachen. In German universities this second degree signifies the major step toward an academic career, the formal authorization to teach. (Most people whose names appear in these pages had doctoral degrees or the degree of Diplom Ingenieur—a sort of advanced master's degree. I will omit titles here, though they were used all the time at Peenemünde.)

Hermann was selected because he was one of the very few aerodynamicists who had both theoretical and experimental experience with supersonic flows. He brought with him from Aachen solid plans for the design of a large supersonic wind tunnel, based on the smaller supersonic tunnel at Carl Wieselsberger's institute. Hermann next assembled a capable group of researchers in fluid dynamics from different universities. When I joined the staff in 1943, the Aerodynamics Institute appeared to have been in existence for a long time, since regular testing took place for two shifts every day. Hermann contributed to the aerodynamic design of the A4 and to later developments. Shortly after the war Hermann disappeared from Kochel in Bavaria, where the wind tunnels were moved after the first major air raid on Peenemünde. At that point he abandoned the primary work of his life for reasons that were obscure to us (see later chapters). He was taken by the U.S. government to Wright Field air base in Ohio, where the new air force began to develop major research laboratories. Later he worked at the University of Minnesota, and finally he moved to the Huntsville campus of the University of Alabama as a professor. He died in 1991 in Huntsville at the age of eighty-six.

Chapter 3: The Supersonic Wind Tunnels

1. See, e.g., *Ernst Mach, Physicist and Philosopher*, ed. Robert S. Cohen and Raymond J. Seeger, Boston Studies in the Philosophy of Science, vol. 6 (Dordrecht, Netherlands: Reidel, 1970).

2. Ludwig Prandtl (1875–1953) began his career at the University of Hannover in 1901. In 1904 he moved to the University of Göttingen as the director of the technical physics department. That same year he proposed the boundary-layer concept, the basis of modern fluid dynamics, which I discuss later. In 1907–1908, while working with a number of leading scientists, he obtained government support for the creation of the first research institute in aerodynamics at Göttingen. With his students, many of whom became great innovators themselves, Prandtl did original work in fluid mechanics and its application to flight. In 1908 he built a low-speed wind tunnel to test his ideas.

In 1911 the German government founded the Kaiser-Wilhelm-Gesellschaft, the forerunner of the Max-Planck-Gesellschaft. A broad range of institutes in many areas of science and engineering sprang up under the umbrella of the newly founded organization. In 1912, Prandtl proposed adding to his new laboratories a Kaiser-Wilhelm-Institute in the fields of aerodynamics and hydrodynamics. It was not until 1925 that what is now the Max-Planck-Institute for Fluid Mechanics was established adjacent to the existing laboratories at the University of Göttingen. An unparalleled complex of facilities covering every basic and applied aspect of the field grew in the 1920s. This initial period of rapid growth included the design and construction of a small supersonic wind tunnel employing a principle of operation that was later used for the Peenemünde tunnels.

Also in the 1920s, Robert A. Millikan (1868–1953), the second American Nobel laureate in physics (1923), became head of the Throop Technical College in Pasadena, California. He renamed the small institution the California Institute of Technology (CalTech) and began to turn it into one of the foremost scientific universities anywhere. Millikan saw the potential importance of southern California as a "nerve center of the nation's aviation industry, requiring an Aerodynamics Institute of scientific advice." (This quote and much of the following material are taken from Michael H. Gorn, *The Universal Man: Theodore von Kármán's Life in Aeronautics* [Washington: Smithsonian Institution Press, 1992].) Daniel Guggenheim, the benefactor of aeronautics at other universities, was approached, and he responded favorably. Prandtl and Theodore von Kármán (1881–1963) were the obvious choices to head the proposed institute. The Guggenheims—father and son—together with Millikan's son Clark, who had studied aerodynamics, suggested inviting Prandtl. From what he had heard about the personalities of Prandtl and von Kármán, Robert Millikan, in a display of remarkable intuition, preferred von Kármán, whom he invited to spend some time (1926–1927) in Pasadena. During this most successful visit it became obvious that von Kármán's wit and urbane personality, together with his scientific abilities, made him the ideal for the California scene and the directorship of the new venture. The Guggenheims were delighted by von Kármán and agreed with Millikan on the appointment. Thus von Kármán joined CalTech as the director of the new Guggenheim Aeronautical Laboratory of the California Institute of Technology (GALCIT), which is still going strong. The work of von Kármán in turn was a major factor in the education of aerodynamicists and in the development of aeronautics in general—including jet propulsion—in the United States. His service as the primary technical adviser to the air force during and after the war was of major importance for the success of American air power.

The Göttingen developments that in part laid the groundwork for the Aerodynamics Institute at Peenemünde have recently been chronicled in Julius C. Rotta,

Die Aerodynamische Versuchsanstalt in Göttingen—ein Werk Ludwig Prandtls (The Aerodynamics Research Laboratory in Göttingen—a work of Ludwig Prandtl) (Göttingen: Vandenhoeck and Ruprecht, 1990). At the time of the seventy-fifth anniversary of the founding of the Göttingen laboratory, Walter Wuest discussed its history in a series of popular articles. These articles have been collected in a book, *Sie zähmten den Sturm* (They mastered the storm), published in 1982 by the DLR (the current abbreviation of the laboratory). The underlying knowledge of fluid dynamics—including some aspects of supersonic flows—accumulated at Prandtl's laboratories and other institutions made possible the rapid developments at Peenemünde.

A more personal biography of Prandtl was recently published by his daughter (Johanna Vogel-Prandtl, *Ludwig Prandtl: Ein Lebensbild, Erinnerungen, Dokumente* (Ludwig Prandtl: A biography, reminiscences, documents) [Göttingen: Max-Planck-Institut for Fluid Mechanics, 1993]). Aside from a discussion of his family and education, it contains interesting details of his views on the Hitler period and the difficulties of research in physics during that period. Prandtl had been attacked by adherents of the Nobel laureates Philipp Lenard and Johannes Stark, who divided physics into "Jewish" and "German" parts. In 1941, Prandtl wrote a long, carefully reasoned letter to Hermann Göring. Extensive enclosures dealt with specific topics in theoretical physics. In this document Prandtl demolished the arguments of the two most politically powerful physicists of the Hitler period. It is remarkable that his own work was, of course, outside the controversial fields of relativity and quantum mechanics. The historian Alan D. Beyerchen (*Scientists under Hitler: Politics and the Physics Community in the Third Reich* [New Haven: Yale University Press, 1988]) cites a number of other examples of Prandtl's attitude. For example, during a dinner sponsored by the German Academy of Aeronautical Research, Prandtl sat next to Heinrich Himmler, the head of the SS. He took up the case of Werner Heisenberg, and he mentioned Einstein's importance. In a later letter he reminded Himmler of this discussion, because the latter had appeared conciliatory at the dinner.

3. In the picture seen on the screen, the degree of grayness is related to the local density in the flow. (Density is defined as mass per unit volume; thus water is denser than air.) In the supersonic flow about a model, the density of the air varies depending on the shape of the model. This optical scheme is therefore an important tool for understanding the flow, supplementing measurements of the forces acting on wind-tunnel models. The optical arrangement is called a schlieren system, from the German word that is now used universally. Schlieren are also visible in the wavering air above a hot radiator or road surface.

4. Pascual Jordan (1902–1980) studied at Göttingen with Max Born. With Werner Heisenberg and Born, he made major contributions to quantum mechanics. In addition, he published in such areas as electrodynamics, cosmology, and relativity. He had the gift of making the achievements of modern physics accessible to a lay audience. A book about the physics of the twentieth century published in 1936 (three years after Hitler came to power) showed his admiration for the work of Einstein, Bohr, Hertz, James Franck, and other Jewish scientists. All of these physicists were attacked under Hitler in the delusion that there was something called "German physics." When I met Jordan he was a professor of theoretical physics at the University of Rostock (a fact that I did not know at the time) temporarily assigned to the Aerodynamics Institute. He was one of a group of conservatives who hoped to influence the regime by a certain degree of cooperation. In 1942 he received

the Max-Planck-Medal, and he moved to the University of Berlin in 1944. After the war he became a professor of philosophy at the University of Hamburg. Later Jordan entered politics as a conservative member of the Christian Democratic Union (CDU), and in 1957–1961 he was one of the few scientists in the West German Bundestag (federal legislature).

5. The boundary layer, or friction layer, is the narrow slice of air along the surface of the entire missile, or in fact the surface of any object moving through air or any other fluid at any speed. Fluids, whether gases or liquids, adhere to surfaces in motion. This can be seen clearly if one looks at the water by the side of a moving ship. In a wind tunnel, where air is blown against a stationary model, the air in the boundary layer comes to a halt at the surface of the model. This effect is observed no matter how well polished the surface may be. This fundamental concept is another contribution by Prandtl.

6. Hermann Kurzweg described the aerodynamics of the A4 in conjunction with the wind-tunnel facilities of Peenemünde in a volume of papers given at an international meeting and published under the aegis of the Advisory Group for Aeronautical Research and Development (AGARD), an aeronautical organization attached to NATO and initially inspired and led by Theodore von Kármán. The volume is *History of German Guided Missiles Development. AGARD First Guided Missiles Seminar, Munich, Germany, April 1956*, ed. Th. Benecke and A. W. Quick (Brunswick, Germany: Verlag Appelhaus, 1957).

7. Wilhelm Keitel (1882–1946), who held the highest rank in the army (*Generalfeldmarschall*), was head of the German armed forces from November 1938 on, after Werner von Blomberg was forced to resign. He was an uncritical follower of Hitler and was not known as a gifted military man. Keitel signed the armistice in 1945, and he was sentenced to death at the Nuremberg trials and subsequently hanged.

8. Mach numbers above 5 are now called hypersonic, with the Greek prefix *hyper* replacing the Latin *super* of the same meaning. For important technical reasons, the regime of Mach numbers above 5 in wind tunnels as well as in the flight of missiles or aircraft, such as the space shuttle, shows major physical differences from the lower supersonic speeds.

Chapter 4: History, People, and Special Events

1. The biographical material on Konstantin Eduardovich Tsiolkovsky and Robert Hutchings Goddard was taken largely from encyclopedias. Most biographies of Hermann Oberth are available only in German. I looked primarily at two books, an older biography by Hans Hartl (*Hermann Oberth* [Hannover: Theodor Oppermann Verlag, 1958]) and what may well be the definitive biography, by Hans Barth (*Hermann Oberth, Werk und Auswirkung auf die spätere Raumfahrtentwicklung* [Work and influence on the later development of space flight] [Feucht: Uni-Verlag Dr. Roth-Oberth, 1985]). The latter book contains much about the history of rocketry and the air raid on Peenemünde. A book by Hermann Oberth himself, which is available in English, may give an idea of the man's way of thinking: *Man into Space: New Projects for Rockets and Space Travel* (New York: Harper, 1957).

2. Michael J. Neufeld, "Weimar Culture and Futuristic Technology: The Rocketry and Space Fad in Germany, 1923–1933," *Technology and Culture* 31 (1990):725–751.

3. Frank H. Winter, *Prelude to the Space Age. The Rocket Societies: 1924–1940* (Washington, D.C.: Smithsonian Institution Press, 1983).

4. To complete the reader's initiation to the Peenemünde army laboratories, I refer to somewhat arbitrarily selected historical works of varying thoroughness. Among them is a book with illustrations of actual wind-tunnel experiments that I worked on.

Here are a few sources of historical information. Some can be found in public libraries, and plenty of additional references can be found in them. The entire corpus of Peenemünde literature is extensive. This is not surprising, considering that the subject still holds interest more than sixty years after the first steps taken at the rocket ports. Most of the writings do not consider the wind tunnels to be particularly important. However, the relation of the aerodynamics research to missile technology is clearly recognized by the writers who actually participated in Peenemünde, starting with von Braun.

Ernst Klee and Otto Merk, *Damals in Peenemünde, ein Dokumentarbericht* (Those days in Peenemünde, a documentary report) (Oldenburg: Gerhard Stalling Verlag, 1963), has extensive illustrations of Peenemünde, including the air force station and most of the late war projects. It also has material on the V2 shots in the American desert after the war. As far as I know, it exists only in German. On p. 77, a photograph shows the Wasserfall being prepared for a test firing. Its shape is based on the drawing reproduced on p. 84, which depicts its final aerodynamic shape. On p. 81 are a drawing and a photograph produced by my group at Kochel of the 1.9-meter-long Taifun antiaircraft rocket, the last military project I was involved with. I have no idea how the authors got hold of this material. One of the main wind-tunnel test sections of the Aerodynamics Institute is shown on p. 115.

A fascinating source of background information on the Hitler period is provided by Robert Wistrich in *Who's Who in Nazi Germany* (London: Weidenfeld and Nicolson, 1982). A German translation was published in Munich by Harnack Verlag (1983). The book contains short biographies—occasionally with photographs—of people working at Peenemünde or involved with rocket development in Berlin or elsewhere, as well as of the many VIPs visiting the place. Moreover, Wistrich includes important people opposed to Hitler's regime. Th. Beneke gives a "Summary of German Developments in Guided Missiles" in the cited AGARD publication *History of German Guided Missiles Development. AGARD First Guided Missiles Seminar, Munich, Germany, April 1956*, ed. Th. Benecke and A. W. Quick (Brunswick, Germany: Verlag Appelhaus, 1957). Of course, Dornberger's two books (chapter 1, note 5) cover many of the events from the earliest days, including such happenings at Peenemünde as the air raid, the A4 production problems, the problems the army had with the SS, and the final flight of the leading people—including von Braun—to Bavaria at the end of the war. David Johnson, *V-1, V-2, Hitler's Vengeance on London* (New York: Stein and Day, 1982), describes the history, including the air raid and the London side of the V1 and V2 bombardments. Finally, here are five readily available books: Wernher von Braun and Frederick I. Ordway III, *Space Travel* (New York: Crowell, 1966); David Baker, *The History and Development of Rocket and Missile Technology* (New York: Crown Publications, 1978); Frederick I. Ordway III and R. Sharpe, *The Rocket Team* (New York: Crowell, 1979); David H. DeVorkin, *Science with a Vengeance* (New York: Springer-Verlag, 1992); Michael J.

Neufeld, *The Rocket and the Reich. Peenemünde and the Coming of the Ballistic Missile Era* (New York: Free Press, 1995).

5. The only discussion devoted solely to the Aerodynamics Institute that I am aware of is by Richard Lehnert, who headed the aerodynamics group in Kurzweg's department, as described in chapter 3. It is a forty-one-page manuscript called "Die Geschichte der wva" (The history of the wva), with the inscription "Silver Spring, Md., 1980." (wva is an abbreviation of the camouflage name of the Aerodynamics Institute relocated in Kochel, Bavaria, following the first British air raid on Peenemünde.) The manuscript contains some interesting material on Lehnert's early days in Peenemünde and on the day that American troops arrived in Kochel. Richard Lehnert kindly gave me permission to use his material.

6. Wernher von Braun (1912–1977) was the outstanding personality who led the Peenemünde team of scientists, engineers, and technicians that developed the A4 missile. Later in the United States, the core group of his team was responsible for many additional achievements in the same field. Most important of this work are two spectacular feats. For one, the development of the carrier rocket on which the first successful satellite of this country, the *Explorer I,* was launched on January 31, 1958. This launching followed closely the first artificial satellite to orbit the earth, *Sputnik I,* which rose from the Soviet Union on October 4, 1957. The climax of the successes of von Braun's group, then at the Marshall Space Flight Center in Huntsville, Alabama, was the development of the huge *Saturn V* rocket carrying the *Apollo 11* spacecraft to the first moon landing by man on July 20, 1969.

There exists, of course, an extensive literature on von Braun's life. The most recent work—proclaimed by its authors, Ernst Stuhlinger and Frederick I. Ordway III, to be *the* biography—*Wernher von Braun, Aufbruch in den Weltraum, die Biographie* (Wernher von Braun, departure into space, the biography) (Esslingen: Bechtle Verlag, 1992). Stuhlinger, a physicist who like me was a soldier assigned to Peenemünde, was one of the leading scientists in von Braun's group both in Germany and in the United States. He now lives in retirement in Huntsville. Ordway, an American, was closely associated with von Braun for twenty-four years. He often appears in the literature as von Braun's coauthor (see note 4 above). The new biography was written in English, and the authors had the usual difficulty finding a publisher in the United States. At that stage the publisher of Dornberger's books (chapter 1, note 5) accepted the manuscript and arranged for a translation into German. After that, the authors found a publisher in the United States, and we now have the original extended two-volume version of the work by Stuhlinger and Ordway, *Wernher von Braun: Crusader for Space* (Melbourne, Fla: Krieger, 1994). The first volume is a 375-page biography, of which only three chapters deal with von Braun's life and work before his departure for the United States. The second volume—called an illustrated memoir—is essentially a collection of photographs and a description of the personal side of von Braun—his parents, his youth, and so forth.

This is the first publication that I have found by a former Peenemünde staff member in which the use of laborers from concentration camps in the Mittelwerk is described in detail. It contains von Braun's impressions of the Mittelwerk, a response to communications from survivors of the camps long after the war, and much additional material on this subject. This two-volume work will most likely not be readily accessible in public libraries, but here are a few books that ought to be available: E.

Bergaust, *Reaching for the Stars* (New York: Doubleday, 1960); D. K. Huzel, *Peene-münde to Canaveral* (Englewood Cliffs, N.J.: Prentice-Hall, 1962); H. B. Walters, *Wernher von Braun—Rocket Pioneer* (New York: Macmillan, 1964); H. M. David, *Wernher von Braun* (New York: Putnam, 1969).

Chapter 5: The British Air Raid

1. David Johnson discusses the air raid and its history, as well as the effects of the V-weapons on London, in his lively book *V-1, V-2, Hitler's Vengeance on London* (New York: Stein and Day, 1982).

2. The raid and its history have appeared in a number of fictional accounts, including James A. Michener, *Space* (London: Corgi Books, 1982). Michener's story has little relation to the actual Peenemünde or—as far as I can tell—the lives of those who followed von Braun to the United States. A well-known fictional account of parts of the Peenemünde story can be found in Thomas Pynchon, *Gravity's Rainbow* (New York: Viking Penguin, 1973).

Chapter 6: The Aftermath and the Move to Bavaria

1. Whenever Dornberger appears, I cite from the books listed in chapter 1, note 5. I did not meet him at Peenemünde, though I often saw him and heard much about him. Near the end of the war I got to know him in Bavaria while taking a car trip with him (chapter 8).

2. I have used the name Peenemünde throughout these reminiscences to refer to the army and air force installations. There is indeed a small fishing village called Peenemünde, and it still exists, as I saw during a visit in 1992. In the course of events, the official name of the army installation frequently changed. Reorganizations took place, but primarily the authorities wanted to confuse the foreign intelligence services. The Army Research Establishment (Heeresversuchsanstalt) was moved on paper to another tiny village called Karlshagen. At one time, it was called *Heimat Artillerie Park* (HAP), or loosely "home artillery depot." Finally (I believe that it was the last name), the quaint title *Elektro-Mechanische Werke* (roughly "electro-mechanical works") survived.

3. I have a copy of an old report whose date I cannot decipher, and I do not remember where I got it. The poorly reproduced document was most likely written at Kochel at the request of one of the groups of scientists in the American army and navy delegations that visited the wind tunnels. I did not hear about the production of this document at the time. It is marked "Restricted," a low level of confidentiality. First, a review of the history of the wind tunnels is given in relation to similar facilities elsewhere. Next, the technical and organizational aspects of the WVA are described in detail, including the names of all participants. The description of the equipment at Kochel refers to the status at the end of the war.

The report is written in poor English, and the full name of the WVA is translated as Hydraulic Experimental Institute Kochelsee G.m.b.H., the letters designating a type of incorporation. Rudolf Hermann is named as the director and chief engineer. (The title of professor is added to his name for reasons unknown to me; he certainly was no professor at Peenemünde.) Two assistant directors are listed: Hermann Kurzweg for science and research and Herbert Graf for commercial operation. Graf was hired to run the business end of the WVA, which had lost the logistic support of the army. I met Graf for the first time after my arrival at Kochel. Finally, Gerhard Eber, who headed the planning of the hypersonic tunnel, is listed as business manager. In fact, however, as far as I know Graf did all the managing of the nonscience side of the installation. The report is of special interest to me because it is the only document in which I find the long-range plans of Hermann and others discussed in some detail.

Planning for the Kochel location of the hypersonic facility—in fact, the first allusion to the separation of aerodynamics from the army laboratory—began in 1942. I was aware of the travel of personnel to Bavaria during my time at Peenemünde prior to the air raid. It is this earlier connection to Kochel that led to the move to the south.

4. Hans Demleitner, whose family has resided in Kochel for several generations, wrote an extensive local history called *Kochel am See*. This large, beautifully bound and illustrated book of 426 pages was published by the author. My brother and I met Demleitner several years ago during one of my visits to Kochel. The mayor, a friend of his, invited the three of us for lunch, where we discussed, among other things, the days of the WVA, to which Demleitner devotes a chapter or so in his book. (The mayor was born after the war.) The first edition of Demleitner's book was criticized, since the author's past under Hitler was obviously that of a staunch follower. This showed in his treatment of Kochel's history under Hitler, and he was forced to withdraw the book. In its purged second edition, the book has an appendix containing copies of newspaper articles whose criticisms of his first version Demleitner strongly disputes.

The author presented me with a copy of the second edition, published by him in 1984. In the book he maintains that a laborer working for a local contractor involved with the WVA claimed that rockets were being built in Kochel. In addition, a sign on one of the freight cars arriving from Peenemünde that should have been removed prior to transport said "Raketen Versuchsstelle Peenemünde" (Rocket Experiment Station Peenemünde). Further rumors concerned the scientists of the WVA, who "would decide the outcome of the war." Unfortunately, Lehnert and I were badly misquoted in Demleitner's opus. Lehnert had regularly visited Kochel after the war on vacation, and he had also met Demleitner.

5. Among the sources on the local level is Richard Lehnert, "Die Geschichte der WVA" (The history of the WVA).

6. Not every experiment was wholly new, of course. Several of the early small supersonic wind tunnels had accumulated a body of technical results on which the work of the Peenemünde tunnels could be based. For example, we saw that Wieselsberger and Hermann had pinned down the fact that driers were needed because the condensation of water vapor spoiled proper testing. At the National Physical Laboratory in England in 1917, the previously mentioned supersonic tunnel whose test section was the size of a thumb permitted a rough determination of the drag of artillery shells based on models the size of a pencil tip. Pressure measurements had been carried out elsewhere before, but no facility had ever come close to providing high-quality flow in such a range of Mach numbers, permitting the use of relatively large models.

The measurement of the drag of a sphere had a remarkable antecedent that must not go unnoticed. Benjamin Robins, a British artillerist and engineer, published sphere drag data in the *Philosophical Transactions of the Royal Society* in 1746–1747.

This remarkable paper was called "Resistance of the Air and Experiments Relating to Air Resistance." The date given is not a misprint. Robins published his work in a period when the colonies still had several years to go before independence from Robins's king. This early aerodynamicist had invented the ballistic pendulum, a hinged board at which he shot spheres, from whose deflection he could infer much about the spheres' flight. He reported sphere drag values at firing speeds up to 1,670 feet/second (509 m/s). E. W. E. Rogers notes that his results agree with modern data (*Aeronautical Journal* 86 [1982]:43). Robins reached M=1.5 with his spheres; however, this fact obviously eluded him, and he reported his puzzling discovery of a substantial increase in drag at certain high speeds. This drag increase appears near the speed of sound, as we now know.

It was more than one hundred years before Ernst Mach showed in 1887 his stunning photographs of bullets in flight that demonstrated shock waves, the source of the increase in drag (*Ernst Mach, Physicist and Philosopher*, ed. Robert S. Cohen and Raymond J. Seeger, Boston Studies in the Philosophy of Science, vol. 6 [Dordrecht, Netherlands: Reidel, 1970]). Again, many years passed before the first international conference on high-speed flight, including supersonic aerodynamics, was held. This seminal gathering of the Volta Conference in Rome in 1935 was attended by almost all researchers in the field, from many countries. A photograph shows all of the conference participants in one compact group, in great contrast to today's mammoth gatherings.

7. In 1950 it was first demonstrated by J. V. Becker at the Langley Laboratory of what is now the National Aeronautics and Space Administration (NASA) that dry wind-tunnel air (i.e., its components, nitrogen and oxygen) does not supersaturate. Consequently, the supply air of a hypersonic wind tunnel must be preheated. A year later, my group at the Naval Ordnance Laboratory in Silver Spring, Maryland, operated the first successful hypersonic wind tunnel in which disturbance-free flows were continuously produced in a large Mach-number range up to M=8.25. This Mach number is about equal to that attempted by Erdmann in 1944.

8. During a discussion at Yale University a number of years ago, the German author Heinrich Böll mentioned that he was a member of a very special group, having been born in 1917 (just like me). That year had the smallest number of births during the first war, and it suffered the largest number of deaths in the second disaster.

9. In Heidelberg we stayed with my wife's aunt. As a minor example of the complex and ambivalent attitudes different people had to the Third Reich, the following may be of interest, though it is unrelated to my narrative. The aunt had been married to a submarine commander who perished late in the first war. Later she remarried, and I met her husband, a retired general of the army who, I believe, had been a divisional commander. He obviously was interested in hearing about my experiences in the armed services. It developed that he was initially a supporter of Hitler on the basis of Hitler's disavowal of the Treaty of Versailles, the rearming of Germany, and the like. Now, however, he was strongly opposed because he could not forgive Hitler for the reckless sacrifice of an entire army at Stalingrad.

Chapter 7: Aerodynamics in the Mountains

1. See chapter 4, note 4.

2. During my study leaves in Berlin, when I helped with the cleanup following

heavy air raids, people spoke freely in the streets, which were brightly lit by flames of the burning buildings. Salvaging belongings was the overriding concern, but many views of the events were loudly proclaimed. Some cursed the attackers, but many made bitter jokes about Hermann Göring, who had promised that all enemy airplanes would be driven off before reaching the capital of the Reich. Even Hitler was accused, and nobody expected anyone to report such reactions to the disastrous air raids. In contrast to Churchill, Hitler never appeared on the streets after an air raid. Of course, he never walked the streets at any time.

3. I looked at archival material after I had written down my own recollections but before I decided to write the book at hand. There arises an eerie feeling when one finds something one had long forgotten. On April 6, 1982, I was comfortably ensconced in the deserted reading room of the archives of the Deutsches Museum on an island of the beautiful Isar River in Munich. The Peenemünde documents were in part chronologically arranged for easy identification. Suddenly I came upon a document on a meeting that took place on September 23, 1944. It was stated that the meeting took place at Kölpiensee, a tiny lake next to Peenemünde where—I am sure—we actually met. The first speaker was listed as Oberleutnant Dr. Wegener, and among the other six participants I knew only Geissler, a mathematician who worked on trajectory calculations. Apparently I brought two top-secret documents, designated 125/44 and 135/44. In addition, I had sketches and diagrams, and with these paraphernalia I reported on the essentially completed aerodynamics of the antiaircraft missile Wasserfall. I also learned that firing experiments of this missile were carried out.

I clearly remember that I visited Berlin in late 1944, and I always wondered how I had gotten there. I had forgotten the meeting at Peenemünde. This is the second—and I hope last—lapse of my memory. I further found in the same folder that Eckert spent December 28–29 in Peenemünde and that Lehnert and Graf met with von Braun on January 25, 1945. The engineers at Peenemünde suggested to Eckert that the external shape of Wasserfall should be used for the projected long-range missile since that configuration had already been worked out. At these meetings, it was stressed that the highest priority ought to be given to this new project. Tests in a continuous supersonic tunnel were demanded, and it had to be pointed out that no such thing existed. There was much additional correspondence between Kurzweg and von Braun about details of the so far unsuccessful experiments with all of the long-range missiles. It is remarkable that such activities still took place quite seriously in late 1944 and early 1945.

4. My father, Paul Wegener (1874–1948), was a famous actor and filmmaker. In Berlin he was one of the stars of the theater of Max Reinhardt, the innovator of a new theatrical style starting prior to World War I and lasting into the 1930s in Germany, when Reinhardt emigrated to the United States. In addition to acting on the stage, my father produced, wrote, directed, and acted in early movies, also starting before 1914. Certain of his films, such as *The Student of Prague* (1919) and *The Golem* (1921 version), are now counted as a part of the classical early cinema, and to this day they are shown in many countries, including the United States. In Berlin my father was known to be highly critical of the Hitler regime, but he was never prohibited from acting. He always thought that Goebbels, who was responsible for most cultural affairs, permitted him a sort of *Narrenfreiheit*, or fool's freedom. His international fame and the fact that many of the great artists of the stage and cinema had left

Germany may have helped him in this respect. He had a wide circle of friends, including academics, writers, and artists, outside the world of the stage. Although he was not himself one of their number, he did know some of the conspirators against Hitler.

Chapter 8: The End of the War

1. There had been attempts to set up production facilities for the A4 at other places. Some of these locations were in the vicinity of suppliers of parts. But the air raids on such places in the industrial areas of Germany forced the decision to go underground and enlarge and equip the Mittelwerk to handle mass production of the missile.

2. Walter Dornberger describes at length the devious moves made by Himmler to obtain absolute power over rocketry (*V2*, trans. James Cleugh and Geoffrey Halliday [New York: Viking, 1954]). He relates a meeting with Himmler at which he carried on—at serious risk to himself—a hopeless fight to retain control over missile development and production.

3. I now know that the "fenced barracks" were actually a concentration camp called by the code name Dora (see chapter 11). Long after the war the German Democratic Republic preserved the few remnants of this camp and turned its grounds into a memorial site. I visited the site in 1992 and 1994, took photographs, and collected literature that had become available since the reunification of Germany. A professional staff at the Dora memorial site is at work on the preservation of files, objects, and the like that shed light on V1 and V2 production.

4. During Hitler's domination of Germany, the discrepancy between what was officially known and what really happened increased with the years. Whenever possible I tried to check for myself the veracity of the rumors that floated around. This was particularly easy during my extended travels to and from the Russian front, travels that could not be pressed into a timetable.

On one occasion that was somewhat similar to but much less frightening than the visit to the Mittelwerk, I entered the ghetto in the large White Russian city of Minsk. I was again visiting my closest friend from student days, whom I had previously seen in Warsaw, as noted in chapter 1. He was now working at an army hospital in Minsk. We borrowed clothing and a primitive horse-drawn sled from Russians that he knew. We could easily enter the loosely fenced area of the extended ghetto. We were struck that some people spoke German, but no German guards were in evidence. We found a lawyer from Frankfurt who told us that most of the inhabitants had been brought there from Germany. This was early in 1942, prior to the mass murders in the large camps. We heard that groups of German Jews were being transported by truck to other places, however, and rumor had it that many had been shot. I remember a crude sign above a door that stated that such and such a person was not to be deported because he was a shoemaker who did fine work for Company X. Here, and in many other instances, German soldiers tried to be helpful. None of us knew about the plans for what came to be known as the Final Solution. But my experiences in the ghetto in Minsk and the V2 factory elicited similar emotions. No more about that.

When I saw my friend in Minsk he was extremely distressed. We spoke about the war, and I told him that I wanted to survive it to build a future as a scientist. He felt that he could not carry on while knowing of the large number of maimed men he was unable to save and of the atrocities committed by Germans against Russians and Jews. With some effort he had himself transferred to serve as a physician with a fighting unit at the northern end of the front. He fell in East Prussia on Christmas Eve in 1944.

Chapter 9: What Next? The Summer of 1945

1. Walter A. McDougall, in his massive 555-page book on the political aspects of the space age (. . . *the Heavens and the Earth, a Political History of the Space Age* [New York: Basic Books, 1985]), for which he received the Pulitzer Prize in history in 1986, notes the extreme rapidity with which news spread that the Peenemünde wind tunnels had been discovered in the "little Bavarian town of Kochel." He claims that only three days after the surrender to the U.S. Army, agents of the Naval Technical Mission were briefed by wind-tunnel people. I was not present at this historic moment when some of our staff (Hermann, Kurzweg?) spoke of artificial space satellites, space stations, and interplanetary voyages. (They had certainly never spoken to me about these topics.) Among the listeners were Clark Millikan and Hsue-shen Tsien. The report of this first group of science people, which certainly preceded the Zwicky report (see note 3 below), "communicated the excitement" to the Navy Bureau of Aeronautics in Washington.

Much later in Pasadena I often saw Millikan, who was Theodore von Kármán's successor as head of the Guggenheim Aeronautical Laboratory at CalTech. I also met Tsien, who had moved from MIT to CalTech. He decided later to return to China, where he played a major role in that country's space efforts. His departure was accompanied by nationwide publicity concerning the possibility of his taking along secret material. No confidential reports or the like were found by a high-level scientific delegation appointed to check this material prior to his departure.

2. Baldur von Schirach (1907–1974) came from an aristocratic family of officers in Berlin (see Robert Wistrich, *Who's Who in Nazi Germany* [London: Weidenfeld and Nicolson, 1982]). The family had artistic and cosmopolitan interests; in fact, his father left the army to become a theater director in Weimar, a job that was once held by Goethe. Von Schirach's mother was an American; two of her ancestors had been signers of the Declaration of Independence. Von Schirach joined the Nazi party in 1925 as a student in Munich, under the influence of the anti-Semitic writings of Houston Stewart Chamberlain. He wrote poetry, and he had great success preaching idealism, bravery, and honor to youth groups. Hitler entrusted his young followers to von Schirach, who in May 1933, at the age of twenty-six, became the youth leader of the Reich. His public appeal was second only to that of Hitler. Through the intrigues of Martin Bormann, who influenced Hitler unseen by the public, von Schirach lost his job in 1940 and joined the army. Later he became the Gauleiter and regent of Austria. At the Nuremberg trials, he was convicted of supporting the deportation of 185,000 Austrian Jews to Poland, and he received a twenty-year sentence. Von Schirach had been one of the very few high Nazi officials who dared to criticize the government, however. In 1943 he had suggested that Hitler treat the conquered people more humanely, and he had criticized the methods of deportation of the Jews. These suggestions alienated Hitler. After he was released from prison in 1966, von Schirach wrote much about the Hitler period, recanted his beliefs, and worked against a rekindling of Nazism in Germany.

I recall von Schirach's life at some length for personal reasons. Since I was born on August 29, 1917, I was fifteen years old when Hitler came to power in January 1933. At my boarding school, politics had not played an active role outside history classes. The school was nonsectarian, and no overt political views were expressed by the teachers. After a long initial period of extreme homesickness, I began to be much attached to the school and its activities, including hiking, travel, and the work we had to perform outside the classroom. Late in the summer of 1933, our director—a staunch German patriot in the respectable, traditional mold, but with modern ideas about education—feared that the school might be taken over by the government and turned into a preparatory school for Nazi party service and the like. (This event indeed happened in 1943–1944, long after I had left.) The director hoped to circumvent a takeover by having all of us join the Hitler Youth. The structure of this organization as led by von Schirach was such that a unit would be just the size of our student body, allowing us to remain independent of the Hitler Youth in the nearby town of Holzminden. My favorite teacher, a Swiss literary scholar, strongly urged me not to follow this idea of the director and to return to Berlin to finish high school. But I was attached to the school, trusted the director, and joined all the students in putting on brown shirts. In fact, this move worked out well: we did what we had always done before, hiking and cycling, camping out, building a sailplane, and the like.

This apolitical branch of the Hitler Youth was particularly unusual for the smallest and youngest of the children, who were entrusted to me by the director. My attitude toward my job was described publicly years later. For the seventy-fifth anniversary of the school in 1984, a book on its history was published. Eberhard Lehmann, a former member of my Hitler Youth group, contributed an article to this volume. Eberhard was four years younger than I. He was the son of the school's director, and after the war he held his father's job for some years. In his article Eberhard says care was taken to appoint group leaders that were older students at the school rather than outsiders. He notes that I was his first group leader and that it was obvious to all that I had not taken the office out of political ambition. He further confirms that I simply continued to do things as before, though we had to wear uniforms.

In September 1933, however, we were briefly involved in the real thing. The Nazi party had planned the first of a yearly series of festivities to praise the harvest and the farmers (*Erntedanktag*). The site of this event was not far from the school, and excepting my little ones, we were ordered to attend. We had to cycle (or walk?) to a field near Bückeburg on the day before the great event and camp in the woods near the site. It rained a great deal, and after the wet night none of us had much enthusiasm left for the happening. We stood at the top of a large meadow that descended away from us, and Adolf Hitler was supposed to speak from a raised stand about a hundred feet away. When an enormous crowd had assembled—it was said that about one million people attended—music and singing started up. Then Hitler gave one of his standard lengthy speeches. Perhaps because I was cold and tired, I did not fall under his spell and was not moved at all. In fact, as the son of an actor who had learned about variety of speech, I disliked the frequent shouting. This attitude never changed, though I saw Hitler on three other occasions in Berlin and was of course exposed to his speeches via radio and newsreels and the newspapers. I simply never understood how anybody could be attracted to him as a person, or to his ideas. My close friends at the university had the same views, in contrast with those of the many young people who were honestly enthusiastic. After I finished high school in the spring of 1936 and moved back to Berlin, I failed to register with the local Hitler Youth as I had been instructed, and nobody approached me. The organization's system of recordkeeping fortunately had its flaws. I never heard von Schirach speak; I wonder whether he might have impressed me at the time. I did look at his poetry, which was not different from that of other minor writers who glorified nature, honor, and so forth.

3. What is now known as the Zwicky report among the few historians who cast a side glance at the wind tunnels is actually the second volume of a document marked "Restricted" and entitled "Report on Certain Phases of War Research in Germany." The report is an excellent summary of the work done at the wind tunnels since their inception, and, understandably, it is better written than the opus by Hermann et al. (chapter 6, note 3). The final version, which includes good line drawings, was prepared at the Aerojet Engineering Corporation in Pasadena. I find myself mentioned three times, once where the organizational structure and the work of different groups are described, and twice in interviews. It is surprising in retrospect that no personal remarks about anyone at the wind tunnels appear in the report. Such remarks were made in other contexts, however, as we shall see. In view of the increasing possibility of taking some of us along, Zwicky's judgment of our widely varying ages, backgrounds, and personalities would be important in a selection. He was certainly the person next to Lieutenant Meyer who knew us best.

Zwicky wrote that cios Team 183 arrived in Kochel to find the equipment intact, and in two weeks the tunnel was made to run. In conclusion, he stated that the "value of the tunnel is such that we proposed to move it to the United States." He also advocated taking fifty German experts along to save several years of work. I find it interesting that Zwicky was apparently the only visitor who made concrete proposals about the institute. His suggestions led to my move to the United States.

Chapter 11: Looking Back

1. Sir Philip B. Joubert de la Ferté, *Rocket* (New York: Philosophical Library, 1957). The "leading astronomer" cited may well have been Fritz Zwicky. In addition, the author describes the August 1943 air raid on Peenemünde in more detail than I have seen elsewhere. Although confusion arose during the raid owing to the artificial fog and actions of less experienced pilots, the pathfinders, as planned, dropped their signal flares near the living quarters rather than on the technical installations. De la Ferté affirms that the raid was the heaviest one on Germany to date in terms of weight of explosives dropped per unit area. He states that the purpose of the raid was to eliminate the people who worked in assembly and development laboratories rather than destroy the installations. As I noted before, I still have difficulty understanding this bombing strategy, which seems particularly inappropriate for a place like Peenemünde, where the technical facilities were concentrated in a small area. De la Ferté provides data on the effects of the V1 and V2 in England. He verifies what I had heard before, that the V1 was psychologically more terrifying than the V2, for the reasons explained to me by my friends in London.

2. The diverse opinions of Hitler held by many people in science (see Alan D. Beyerchen, *Scientists under Hitler: Politics and the Physics Community in the Third*

Reich [New Haven: Yale University Press, 1988]) were surely also found at Peenemünde. Interesting additional material is provided by Karlheinz Ludwig, *Technik und Ingenieure im dritten Reich* (Technology and engineers in the Third Reich) (Düsseldorf: Droste Verlag, 1974). Ludwig looks at the actions of professional societies, such as the well-known VDI, the Verein Deutscher Ingenieure (Society of German Engineers), during the Hitler period. He finds that engineers generally gave up their independence much more rapidly than scientists such as physicists. He does not have much to say about Peenemünde.

3. James McGovern, *Crossbow and Overcast* (New York: Paperback Library, 1966); Clarence G. Lasby, *Project Paperclip, German Scientists and the Cold War* (New York: Athenaeum, 1971); Tom Bower, *The Paperclip Conspiracy* (London: Michael Joseph, 1987); Linda Hunt, *Secret Agenda: The United States Government, Nazi Scientists, and Project Paperclip, 1945 to 1990* (New York: St. Martin's Press, 1991); Marsha Freeman, *How We Got to the Moon* (Washington, D.C.: 21st Century Science Associates, 1993). The preceding books deal primarily with Project Paperclip. The exception is Freeman's work, in which only one chapter is devoted to the subject. In addition, some of the sources previously listed as dealing with Peenemünde and the A4 also discuss Paperclip. For example, Ernst Stuhlinger and Frederick I. Ordway report specifically on the adventures of the von Braun group, the largest single group that collectively moved to the United States (*Wernher von Braun: Crusader for Space* [Melbourne, Fla.: Krieger, 1994]). Walter A. McDougall writes about the reconstruction of the wind tunnels, a subject neglected by most discussions of Peenemünde (. . . *the Heavens and the Earth, a Political History of the Space Age* [New York: Basic Books, 1985]). David H. DeVorkin describes how "the V2 travels west" (*Science with a Vengeance* [New York: Springer-Verlag, 1992]).

4. Two books written by participants in the rocketry work performed by German scientists and technicians in the Soviet Union were recently published: Werner Albring, *Gorodomlia. Deutsche Raketenforscher in Russland* (Gorodomlia. German rocket scientists in Russia) (Hamburg: Luchterhand, 1991) (Gorodomlia is the island in the river Volga between Moscow and St. Petersburg [formerly Leningrad] to which the Germans were confined with their families); and Kurt Magnus, *Raketensklaven. Deutsche Forscher hinter rotem Stacheldraht* (Rocket slaves. German scientists behind red barbed wire) (Stuttgart: Deutsche Verlagsanstalt, 1993). James McGovern (see previous note) discusses the events surrounding the V2 after the war in the Soviet zone of Germany in a captivating style and provides additional earlier sources. Finally, Gröttrup's wife wrote her story in English (Irmgard Gröttrup, *Rocket Wife* [London: A. Deutsche, 1995]); unfortunately, I was unable to find the book in the United States or in London.

5. Edward Pachaly and Kurt Pelny, *Konzentrationslager Mittelbau-Dora* (Concentration camp Mittelbau-Dora) (Berlin: Dietz Verlag, 1990); Udo Breger, *The Rocket Mountain* (Ostheim/Rhön: Verlag Peter Engstler, 1992); Rainer Eisfeld, *Die unmenschliche Fabrik: V2-Produktion und KZ Mittelbau Dora* (The inhuman factory: V2 production and KZ Mittelbau Dora) (Nordhausen: KZ-Gedenkstätte Mittelbau-Dora, 1992); Angela Fiedermann, Torsten Hess, and Markus Jaeger, *Das Konzentrationslager Mittelbau Dora* (The concentration camp Mittelbau Dora) (Berlin: Westkreuz Verlag, 1994); Joachim Neander, *Die Letzten von Dora* (The last of Dora) (Berlin: Westkreuz Verlag, 1994); Ulrich Brunzel, *Hitlers Geheimobjekte*

in Thüringen (Hitler's secret objects in Thuringia) (Meiningen: Heinrich-Jung-Verlagsgesellschaft, 1994).

My current knowledge of the history of the Mittelwerk and the associated concentration camps is based on the preceding references, correspondence and discussions with German historians, and visits in 1992 and 1994 to the memorial site Mittelwerk-Dora in Nordhausen. Torsten Hess was my guide during my second visit. A historian of technology, he is working on preservation and documentation at the memorial site. In addition, he organizes meetings of former prisoners from several countries, which take place periodically at Nordhausen. Progress in conservation was made between my visits. A professional staff was appointed by the state government, documents were collected, and the underground factory was successfully preserved. The state of Thuringia had declared the Kohnstein and the site of the camp Dora to be landmarks. This kept a mining company from turning the whole mountain into building materials, disregarding the possibility that dead prisoners might still be found in the tunnel system. Unfortunately, I was not permitted to enter the sealed V2 production tunnels in 1994, owing to insurance problems.

In addition to the cited literature, discussions, and correspondence with historians mentioned in the acknowledgments, I relied on the statements by von Braun recorded in the Stuhlinger and Ordway biography. Moreover, in this biography much of the older literature on the Mittelwerk and related subjects is cited. This includes postwar books and reports by former prisoners. A recent book (Michael J. Neufeld, *The Rocket and the Reich. Peenemünde and the Coming of the Ballistic Missile Era* [New York: Free Press, 1995]) supplies in part newly discovered archival material on the subject of the Mittelwerk, the associated concentration camps, and the German staff of scientists and technicians who worked in and around the Mittelwerk, and who had moved from Peenemünde to the Nordhausen area. Again, an extensive bibliography appears.

6. Some of the books cited earlier (Pachaly and Pelny, Eisfeld, and Fiedermann, Hess, and Jaeger) display facsimiles of important documents. Additional copies of original letters, memorandums, and other items were graciously provided by Rainer Eisfeld of the University of Osnabrück (Germany). They are listed in chronological order: (a) April 16, 1943. Memorandum on Rudolph's visit to the Heinkel works on April 12, 1943, signed by him and proposing use of concentration camp labor at Peenemünde. (b) August 25, 1943. Minutes of a meeting of seven members of the Peenemünde staff, including von Braun, on the distribution of V2 work among several contractors. One firm proposed to use concentration camp labor. (c) June 22, 1944. A stern directive from the civilian directors to the department heads at the Mittelwerk to follow the orders of the SS prohibiting physical attacks by the civilian staff on the concentration camp inmates. (d) August 15, 1944. Letter from von Braun to A. Sawatzki (a director of the Mittelwerk) proposing improved living conditions for a French professor selected by von Braun at the concentration camp Buchenwald for work on the V2 assembly. (e) May 1, 1992. The judgment of the Federal Court of Appeal of Canada in Ottawa, Ontario, Canada, giving the reason for the denial of the application for Canadian citizenship by Arthur Rudolph.

7. One example of an archival report that did not reflect reality is given by my discovery of the minutes of the meeting on Wasserfall held on September 23, 1944 (described in chapter 7). I had forgotten this strange conference and did not receive a

copy of the minutes. At Peenemünde seven people sat around a table and discussed the status of the Wasserfall. I was the first speaker. It is purely coincidental that what I had to say was the truth—that the aerodynamics of the missile had been in hand for some time. Consequently its external shape was defined. I am reasonably sure that everyone around that table knew that the actual completion and manufacture of this complicated device could not possibly be expected in the near future, and that the war was lost. But by reading the minutes of this gathering now, one might easily assume that a group of believers in the final victory expected the antiaircraft rocket to rise into the sky to destroy enemy airplanes. I do not propose to compare the account of this simple technical meeting with, say, von Braun's letter about Buchenwald, a letter that poses more profound questions, none of which can be answered with certainty at this time.

Index